An Introduction to Chaotic Dynamics

Classical and quantum

Online at: https://doi.org/10.1088/978-0-7503-6453-9

An Introduction to Chaotic Dynamics

Classical and quantum

Robert C Bishop

Physics and Engineering Department, Wheaton College, IL, USA

IOP Publishing, Bristol, UK

ISBN   978-0-7503-6453-9 (ebook)
ISBN   978-0-7503-6456-0 (print)
ISBN   978-0-7503-6455-3 (myPrint)
ISBN   978-0-7503-6454-6 (mobi)

DOI   10.1088/978-0-7503-6453-9

Version: 20250501

IOP ebooks

British Library Cataloguing-in-Publication Data: A catalogue record for this book is available from the British Library.

Published by IOP Publishing, wholly owned by The Institute of Physics, London

IOP Publishing, No.2 The Distillery, Glassfields, Avon Street, Bristol, BS2 0GR, UK

US Office: IOP Publishing, Inc., 190 North Independence Mall West, Suite 601, Philadelphia, PA 19106, USA

To my dad, Emerson Edward (Ed) Bishop, whose warm smile will be missed.

Contents

Preface

Chaotic dynamics, more commonly known as chaos, has captured scientific as well as popular imagination since the 1960s. Much work has been been done on chaos over the decades, all of which is fascinating from a mathematics and physics perspective, but which also has actual-world implications and applications. The realization that nonlinear systems are ubiquitous in nature, that 'noise' and irregular behavior are as if not more important than regular, repeatable linear behavior, has revolutionized how scientists think about the world.

This book is intended as an introduction to chaotic dynamics (What is it? How does it arise? Why is it important?) easily accessible for mathematics and physics undergraduate and beginning postgraduate students who have had a first course in mechanics and quantum mechanics. Yet, it also should be accessible for readers with an interest in mathematics or physics who have a mathematical background through differential equations and basic set theory.

I begin by introducing preliminary concepts important for understanding the literature on chaotic dynamics as well as a brief history of chaos to situate research in the flow of 20th-century developments in mathematics and physics. The phenomenology of chaotic behavior is then described for both conservative and dissipative dynamical systems illustrating the intricate order of chaotic dynamics.

Rarely is the problem of defining chaos discussed in the math and physics literature in any detail. It is often thought of as a more 'philosophical' problem, but has import for how we think about what chaotic dynamics is. In particular, the issue of defining chaos comes to a head in the question of quantum chaos. I devote a chapter to the issues of definition and propose yet another attempt at covering the number of cases that defy existing definitions.

I then turn to implications chaotic dynamics has for modeling actual-world systems and predicting their behaviors followed by chapters discussing quantum chaos and its relationship to macroscopic chaotic systems (Can chaos amplify quantum effects undermining determinism of macroscopic-scale systems? Is quantum chaos the same as classical chaos? What is the status of the correspondence principle and the relationship between quantum and classical physics?).

The book rounds out with a discussion of applications and broader implications of chaos in our world for determinism, laws, causation, reduction and emergence, free will and divine action in the world. The latter topics have been long-standing interests of mine since at least my PhD days. Chaos and nonlinear dynamics revamp much of our thinking about these subjects, given how many of our intuitions are honed on linear dynamics and simple mathematical and physical models. Our intuitions are in for some surprises.

My hope is this book provides an enjoyable and insightful introduction to chaos suitable as a supplement for a one-semester course on the technical details and examples of chaotic dynamics, but also for those interested in learning about and exploring classical and quantum chaos and their implications for our world.

Acknowledgements

No book is possible without the help of others through conversations, interacting with their works, dealing with referee reports, and the like. Over the years, discussions with Harald Atmanspacher, Hans Primas, Michael Silberstein, and Leonard Smith, among others, have been fruitful and shaping of my journey with nonlinear dynamics and chaos. The early path for this journey was set by my dissertation work on chaos, indeterminism, and free will under the direction of Robert Kane and Fred Kronz. I learned much about how to situate physics in a larger perspective from undergraduate and masters theses under John Archibald Wheeler, an experience that has shaped my somewhat chaotic journey. Of course, Phoebe Hooper, Chris Benson and their colleagues at Institute of Physics Publishing have been very helpful all along the way making the process of this book's emergence as smooth as I could imagine.

It is difficult to assign the relative weight of the influences each person has had on what has led up to a rather unique book on the subject of chaos, but I am thankful to them all.

Author biography

Robert C Bishop

Robert C Bishop is Professor of Physics and Philosophy and the John and Madeleine MacIntyre Endowed Professor of Philosophy and History of Science in the Physics and Engineering Department at Wheaton College. His research focuses on history and philosophy of physics and the social sciences and free will, with special attention to emergence, determinism, chaos, and complexity. He is the author of *The Physics of Emergence*, Second Edition (Institute of Physics Press, 2024), and *Chaos Theory: A Quick Immersion* (Tibidabo Publishing, 2023), and co-author of *Emergence in Context: A Science-First Approach to Metaphysics* (Oxford University Press, 2022).

IOP Publishing

An Introduction to Chaotic Dynamics
Classical and quantum
Robert C Bishop

Chapter 1

Preliminaries

This chapter lays out preliminary material needed to understand the literature on chaos and provides background for the rest of this book. Clarifying what scientists mean by concepts such as randomness, uncertainty, and sensitive dependence, along with introducing dynamical systems and related concepts are important for understanding chaotic dynamics and the relationship between chaos in our mathematical models and phenomena in the actual world.

1.1 Randomness

It is important to begin with a confusion about randomness or random behavior since chaotic behavior is often described as 'random'. In everyday language, the word random is often used to mean lawless or unordered behavior, lacking any pattern or structure. In the sciences, however, there are no known examples of such lawless disorder in any of the physical phenomena we study.

The kinds of randomness scientists study require disambiguation, though. One form of randomness is **apparent randomness**, where a system's behavior lacks any apparent pattern when we observe it but has an underlying deterministic order. An example would be the role of dice. Outcomes of dice rolls appear to have no order or pattern. But suppose we were able to know the precise velocity of the dice as they leave the hand, the friction of the table they land on, the exactness of the square shape of the dice, the deviations from uniformity in their density, among several other factors. Given these factors, the faces turned upwards when the dice come to rest are fully determined. We cannot calculate the outcome of the throw due to the many factors involved and the limits on our epistemic access to these factors, but there is an underlying deterministic order. The randomness is only apparent because of our limited knowledge: the behavior of the dice is fully determined in an ordered way. Yet, in the long run statistical patterns emerge. For instance, with enough rolls more sevens will be the outcome than any other total number from the upturned

doi:10.1088/978-0-7503-6453-9ch1

1-1

faces because there are more combinations of dice faces adding up to seven. Similarly, snake eyes and box cars will be the least common statistically.

The second form of randomness scientists study is known as **irreducible randomness**, where the full set of physical conditions determine the probability for outcomes, but not the specific outcome in a system at a particular time. Nonetheless, the irreducible randomness of these outcomes conforms to fixed probabilities. These probabilities are characterized by statistical laws rather than deterministic laws in contrast to apparent randomness. Irreducible randomness is a different form of order than the deterministic order we experience with mechanical systems such as engines and computers, but it's not lawless chaos. As in the apparent randomness case, persistent statistical patterns emerge though, in contrast, not because there is any underlying deterministic mechanism.

Radioactive decay is an example of irreducible randomness. All the relevant factors in a sample of radium will not determine when any specific nucleus in the sample undergoes a decay event. Nonetheless, the sample will behave as described by a statistical law characterizing how many nuclei will decay on average during a given time interval. We make use of such irreducible randomness all the time in nuclear power, medical treatments, and so forth.

Being clear about the type of randomness exhibited by a system's behaving chaotically avoids common misconceptions that such systems lack order. Chaos actually exhibits intricate order appearing random at first blush. More will be said about how measures for randomness might be made more precise in Appendix A.

1.2 Uncertainty

Uncertainty is another important concept for understanding chaos. In the sciences, uncertainty is the lack of exactness, accuracy, or precision in a measurement. Such uncertainty can be introduced into a measurement of the state of a system through limitations on the accuracy or sensitivity of a measuring apparatus, or through some source of fluctuations in the state of the system being measured, or through the act of measurement introducing a small disturbance into the state being measured by the measurement apparatus, among other possibilities.

There is nothing new about imprecision and uncertainty in measurements. For example, Pierre Duhem (1861–1916) [1] warned about the imprecision of measurement apparatus (among other forms of uncertainty in experiments) in the early 20th century. Unlike some philosophical analyses, scientists rarely assume that their observations and measurement apparatus can be made as arbitrarily accurate as is needed for any argument to succeed.

Think of uncertainty as represented by a small volume around the measured value of the initial state or conditions of the system. Scientists represent such uncertainty by reporting measurements with error bars, specifying the range of uncertainty within which a measured value lies (e.g., a person's height being measured as $1.68 \pm .021$ m). The uncertainty in a measurement of the state of a system can grow over time in mathematical models as reflected by a forecast for the system's future behavior. For instance, in their forecasts of tropical storms and hurricanes, the National Oceanic and

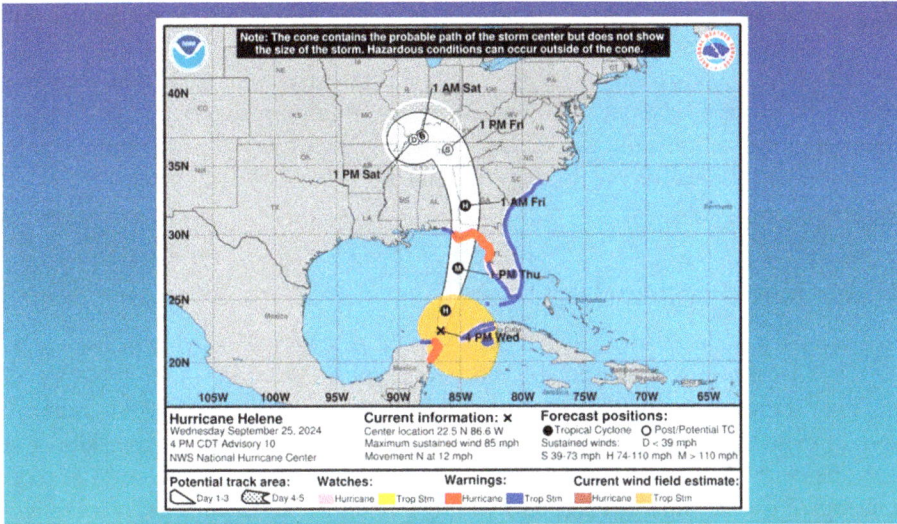

Figure 1.1. Forecast for hurricane Helene. Uncertainty in the initial conditions spreads rapidly leading to a large cone of possible paths the storm could take. Courtesy of the National Oceanic and Atmospheric Administration (NOAA).

Atmospheric Administration (NOAA) uses a cone to represent the growth in uncertainty in the storm's position, speed, and other factors from the initial measurements as in figure 1.1.

This figure is from a computer forecast for hurricane Helene generated on 25 September 2024 at 4 pm while still in the Gulf of Mexico. The measurement imprecision in the starting conditions for Helene's origin results in uncertainty growing rapidly leading to a large cone representing the possible paths the storm could take. It is important to keep in mind that the cone doesn't represent the increasing size of the storm but the possible range of paths the center of the storm could take. The forecast in figure 1.1 represents the results of many modeling techniques designed to mitigate the growth in the uncertainty in the initial conditions, otherwise the spreading of the cone would be even larger (more on this later).

One important feature of chaos discussed in the literature is the rapid growth of uncertainty due to limitations on epistemic access to the precise state of systems typically called sensitive dependence on initial conditions (section 1.8).

1.3 Determinism

A system (e.g., a set of equations defining a model), is deterministic if it exhibits unique evolution:

Definition 1.1. Unique evolution: a given state of a model is always followed by the same history of state transitions. More precisely, a model $m \in M$ exhibits unique evolution if every model $m' \in M$ isomorphic to m with respect to the set of relevant laws L undergoes the same evolution as m [2].

Chaotic behavior is always deterministic, though sometimes equating determinism with predictability leads to confusion about this point. Another confusion about determinism and chaos can arise if the state space (section 1.5) used to analyze chaotic behavior is coarse-grained. Such coarse graining produces an epistemic form of nondeterminism but the underlying dynamical equations still exhibit unique evolution. Possibilities for breakdown of determinism in chaotic systems is explored in section 10.3.1.

1.4 Dynamical systems[1]

Chaos typically is understood as a mathematical property of a **dynamical system**. A dynamical system is a deterministic mathematical model describing system evolution. Time can be either discrete or continuous. A one-dimensional dynamical system is called a map.[2] For a simple example of a discrete map, consider a rule where one penny is placed on the first square of a checkerboard, two on the second square, three on the third square, and so forth. The output or number of pennies on each square represents a step in time. This is an iterative process: the current value of an observable at time t, the number of pennies, has the rule applied to it to yield the value of the observable at the next time step $t + 1$. Then the rule is applied to this value to produce a new output at time step $t + 2$ and so on. This procedure builds up a time series of outputs from the map for each time step. The number of pennies on square one (t), square two ($t + 1$), square three ($t + 2$) and so on would be an example of a discrete time series produced by iterating a map. Daily hospitalization rates for COVID-19 would be another example of a time series.

Dynamical systems of two dimensions or higher typically are called flows. Figure 1.1 is an example of a flow, where the mathematical model of the developing hurricane is multidimensional. Flows can also be discrete or continuous and their outputs also form a time series albeit of two or more dimensions.

Such models may be studied as mathematical objects or used to describe target systems (e.g., a physical, biological, or economic system). A simple example would be the equation describing the motion of a pendulum. The equations of a dynamical system are often referred to as dynamical or evolution equations describing the change in time of observables taken to adequately describe the target system (e.g., momentum and position as functions of time for a pendulum). A complete specification of the initial state of such equations is referred to as the **initial conditions** for the model, while a characterization of the boundaries for the model domain are known as the **boundary conditions**. An example of a dynamical system with a boundary condition would be the equation modeling the flight of a rubber ball thrown against a wall. The boundary condition might be that the wall absorbs no kinetic energy (energy of motion) so that the ball is reflected off the wall with no loss of energy. The initial conditions would be the position and velocity of the ball as

---

[1] This section draws heavily from [3].
[2] Sometimes the literature deviates from this usage, as in the two-dimensional baker's map.

it left the hand. The dynamical system would then describe the flight of the ball to and from the wall under these conditions.

1.5 State space

Scientists mathematically model of target systems using an abstract mathematical space of points called **state space**, where each point represents a possible state of the system.[3] A state is assumed to be characterized by the instantaneous magnitudes of observables relevant for a complete description of system states and behaviors. The dynamical equation describes how a system behaves in state space. The number of independent observables needed to characterize the system state determines the number of dimensions of the relevant state space. For instance, the independent observables characterizing a one-dimensional pendulum are its position and momentum, x and p_x, so the relevant state space (phase space) has dimension two.

The equations in the dynamical system usually have one or more parameters determining the strength of the contributions of particular terms in the mathematical model to the system's behavior in state space. As we will see, the initial state and parameter magnitudes in the mathematical model lead directly to the emergence of chaos for nonlinear (section 1.7) equations.

For either a map or flow, the initial state is evolved by the dynamical system producing a trajectory in state space representing a time series of state transitions. Trajectories are often called orbits (particularly when they exhibit periodic behavior). It is often the case that useful geometric properties of the dynamical system can be studied in state space without explicitly solving the equations.

1.6 Attractors

One important property of dynamical systems is that of an **attractor**. An attractor is the value an iterated map returns in state space (or set of values in the case of a flow). System trajectories converge on the same point repeatedly or get arbitrarily close to a specific point in state space. Attractors are defined over the smallest set of points such that an attractor cannot be decomposed into two or more attractors with distinct basins of attraction.

There are four main kinds of attractors:
1. **Fixed points (point attractors):** dynamical systems produce the same output (e.g., the resting position $x = 0$ of a damped pendulum).
2. **Periodic loops (limit cycle attractors):** dynamical systems periodically repeat the same sequence of outputs in a regular pattern (e.g., a perfectly driven pendulum).
3. **Quasi-periodic loops:** dynamical systems have more than one period but with no fixed period (e.g., tides rolling in and out) or exhibit behavior with two or

---

[3] A state characterized by position and momentum observables is often called **phase space**, though some authors use the term phase space for any state space.

more periods of incommensurable frequencies (e.g., the van der Pol equation describing a vacuum tube oscillating circuit[4]).

4. **Aperiodic:** dynamical system outputs appear to randomly jump around but intricate order exists (the chaos we seek lurks here).[5]

When the outputs of a dynamical system approach some attractor, mathematicians say that the system is **dissipative**. A dissipative dynamical system has the property of shrinking its activity to a smaller area or volume of state space. This is to say that dissipative systems don't preserve the state space volume of their activity. Moreover, dissipative systems don't conserve energy. Such systems are also called **nonconservative**. Observing such contracting behavior guarantees that an attractor exists in that region of state space. A dynamical system whose activity doesn't shrink to a smaller volume in state space is called **Hamiltonian**. Such systems are also known as **conservative** because they conserve the state space volume of the activity. Furthermore, Hamiltonian systems conserve energy. Chaotic dynamics occurs in both dissipative and Hamiltonian systems.

1.7 Nonlinear systems

Dynamical systems of interest in chaos studies are nonlinear, such as the Lorenz model equations for convection in fluids [4]:

$$\frac{dX}{dt} = -\sigma X + \sigma Y \tag{1.1}$$

$$\frac{dY}{dt} = rX - Y - XZ \tag{1.2}$$

$$\frac{dZ}{dt} = XY - bZ \tag{1.3}$$

where X is proportional the circulatory fluid flow, Y is proportional to the temperature difference between rising and falling fluid regions, and Z is proportional to the distortion of the vertical temperature profile from linearity, with σ the Prandtl number (kinematic viscosity divided by thermal conductivity), r the Rayleigh number (a measure of the relative strength of convection to diffusion) divided by its critical value, and b a constant.

Dynamical systems are characterized as linear or nonlinear depending on the nature of the equations. Consider a differential equation system $d\vec{x}/dt = F\vec{x}$ for a

---

[4] This is a second-order differential equation:

$$\frac{d^2y}{dt^2} + \mu(1 - y^2)\frac{dy}{dt} + y = 0.$$

When $\mu = 0$ the van der Pol equation reduces to an equation for simple harmonic motion, another example of a limit cycle or loop attractor.

[5] A fifth kind of attractor, known as a **strange attractor** is discussed in section 4.1.

set of observables $\vec{x} = x_1, x_2, \ldots, x_n$. These observables might represent positions, momenta, or other properties of a target system, and the system of equations describes how these key observables change over time. The Lorenz system is an example.

Let $x_1(t)$ and $x_2(t)$ be solutions of the equation system system $d\vec{x}/dt = F\vec{x}$. When the system of equations is linear, it can easily be shown that $x_3(t) = c_1 x_1(t) + c_2 x_2(t)$ also is a solution, where c_1 and c_2 are arbitrary constants. This is the **principle of linear superposition**. If this principle holds, then a system behaves linearly: any multiplicative change in an observable by a factor α implies a multiplicative or proportional change of its output by α.

In a nonlinear system, such as Lorenz, linear superposition fails, and a system need not change proportionally to the change in a observable or a parameter such as the Rayleigh number.

1.8 Sensitive dependence

One of the most exciting ideas about chaos is that of very small changes now producing very large effects in the future. The illustration of butterfly wing flaps in Brazil causing a tornado in Texas three weeks later is gripping. A butterfly wing flap disturbs a relatively small number of air molecules. Amplifying such a small disturbance into a tornado thousands of kilometers away suggests an exquisite sensitivity to such minute disturbances. This phenomenon is called **sensitive dependence on initial conditions** (SDIC), but this sensitivity must be qualified more precisely to capture one of the most important features of chaotic dynamics.

Consider a car at a stop light. When the light turns green, if you press the accelerator a little bit, the car inches forward slowly. Press the accelerator a little more and the car increases speed a little more. The small changes in the accelerator do not lead to rapid increases in the car's speed as these are linear responses.

Now consider a perfectly symmetric cone precisely balanced on its tip with only the force of gravity acting on it. A perfectly balanced cone would maintain this unstable equilibrium forever in the absence of disturbances. However, in the actual world, the smallest nudge from an air molecule colliding with the cone will cause it to tip over. Yet, the cone could tip over in any direction due to the slight differences in various perturbations arising from collisions with different molecules. This example illustrates that variations in the slightest of causes produce dramatically different effects. Plotting the tipping over of the unstable cone in state space would reveal that from a small ball of potential starting conditions representing the variations in air molecule collisions (apparent randomness), several different directions for the cone's falling issue forth from this small ball of uncertainty. Dramatic, indeed, but not what scientists mean by the SDIC characteristic of chaotic dynamics.

Instead, exponential growth in uncertainties is usually recognized as the crucial mark for chaos. Sensitive dependence implies that any uncertainty in the initial state of a system can rapidly lead to very large uncertainty about the system's behavior when the uncertainty grows exponentially with time.

For example, measurement of an initial state of a system yields some degree of uncertainty. Under a linear map, reducing the initial uncertainty by a factor of one hundred implies the evolution of the system can be forecast one hundred times longer into the future before the growth in uncertainty exceeds this factor of one hundred. By contrast, under an exponential map, reducing the initial uncertainty by one hundred only allows accurate forecasts of future behavior four times as long into the future.

Exponential growth in uncertainties creates challenges for forecasting the future of chaotic dynamics. It isn't possible to reduce uncertainty to zero in our measurement of initial conditions due to mechanical and other limitations in our observations—there are no perfect measurements. Hence, SDIC in our models has actual-world consequences for forecasting the future for systems behaving chaotically.

Consider weather forecasting. Henri Poincaré (1854–1912) noted that cyclone behavior was sensitive to the precise location where they originate:

> We see that great perturbations generally happen in regions where the atmosphere is in unstable equilibrium. The meteorologists are aware that this equilibrium is unstable, that a cyclone is arising somewhere; but where they can not tell; one-tenth of a degree more or less at any point, and the cyclone bursts here and not there, and spreads its ravages over countries it would have spared. This we could have foreseen if we had known that tenth of a degree, but the observations were neither sufficiently close nor sufficiently precise, ... Here again we find the same contrast between a very slight cause, unappreciable to the observer, and important effects, which are sometimes tremendous disasters ([5], p 398).

This illustrates the intersection between apparent randomness and sensitive dependence. Observations of the origin point aren't precise enough, so cyclones 'burst' on the scene with their ravages in a manner appearing random. Although Poincaré was unaware of the modern notion of SDIC, uncertainty in the precise origin of cyclones is the same as the case where a relatively small ball of uncertainties grows exponentially rapidly (compare with figure 1.1). Modern meteorology, is greatly improved in forecasting skill relative to the early 20th century but still must contend with sensitive dependence due to chaotic dynamics and uncertainty in our data (more on this later).

One aid to making forecasts under uncertainty is that exponential growth doesn't go on forever. The finitude of physical space or other resources implies growth in uncertainty saturates at a finite amount. Return to the example of pennies stacked on a checkerboard. Restricting ourselves to US pennies, it is estimated that about 200 billion are currently in circulation. If the stacking algorithm is exponential, the growth in the number of pennies would halt at approximately square 21. Even adding in all the British pence and pennies from all other countries, there aren't enough pennies in the world to support growth to square 22.

If the growth of uncertainty representing the divergence of trajectories issuing forth from nearby points can't go on forever, what happens? The nonlinearities in our model dynamical systems cause neighboring trajectories to diverge away from each other rapidly as time advances. If the trajectories are to avoid leaving state space, then they must fold back into that space. It is the stretching and folding of trajectories that is a key indicator of nonlinear dynamics and, under the right conditions, chaos. The stretching of trajectories is associated with the explosive growth in uncertainties while the folding confines all trajectories to a region of state space. Think of the stretching and folding of taffy in a rotating taffy pulling machine [6]. This kind of dynamics allows conservative dynamical systems to preserve state space volume as the dynamics plays out over time. It is this stretching and folding of trajectories that is a main contributor to SDIC in chaotic dynamics.

1.9 Models and faithfulness[6]

Mathematical models are simplified descriptions based on key observables and processes of a system under study. Simplified is relative to context, however. For instance, models are constantly used in economics and public policy, public health (e.g., in the COVID-19 pandemic), weather forecasting, and in anthropogenic global warming leading to climate change, among other contexts. Although scientists cannot include every possible factor in these models, observables and parameters considered most important are included as well as the processes deemed most relevant. Policy makers use such models to make decisions affecting our daily lives as well as those of future generations. Therefore, it's important to understand what we're doing when we're using models for understanding nature's working and as guides to decision making.

There is a close connection between models used to produce a weather or sunspot forecast and the state space of observables thought important to track (section 1.5). Nevertheless, note the assumptions being made here that often pass unexamined: scientists assume the state of the target system of interest—a weather system or the Sun, say—is appropriately characterized by the magnitudes of the state space observables. They further assume that the system state represented in state space corresponds accurately to the state of an actual-world system through these magnitudes. Furthermore, scientists assume the model equations capture actual processes relevant to the target system as well as that the state space accurately represents all the possibilities for the target system's behavior.

These assumptions allow the development of mathematical models for the evolution of system states in state space and such models are taken to faithfully represent the target system. This the **faithful model framework**. If the equations faithfully represent important features of the target system, and the state space faithfully represents the target system's possible behavior, then one can reasonably think of plotting trajectories for both the target system and its model in the same state space. Later, we'll raise some questions about how faithfully we can represent

---

[6] This section draws heavily from [3].

actual-world target systems in the state space when using computers. But here are some delicate modeling issues that will be with us throughout the book.

In its idealized form, the faithful model framework becomes the **perfect model framework**: mathematical models are taken to be perfect representations of target systems. Often it's the case that perfect models are assumed so that analysis is implicitly carried out within the perfect model framework even though our models may be flawed in various ways.

One implication of the perfect model framework is that it licenses (often sloppy) switching back and forth between model talk and target system talk. For instance, if a weather model is presumed to be a perfect representation of the Earth's weather system, then any statements made about the model should also be true of the weather being modeled. Nonetheless, using the perfect model framework when we do not have perfect models is problematic, particularly in contexts where we are dealing with nonlinear effects.

In reality, scientists have neither perfect models nor perfect data. Instead, scientists generally end up following piecemeal strategies for improving our models and data for actual-world systems (representing competing approaches vying for government funding).[7]

The first basic approach focuses on piecemeal improvements to data accuracy while keeping the model fixed. The intuition is that so long as a model is faithfully reproducing target system behavior, improving data precision will lead to model output monotonically converging to target system behavior more accurately.

The second basic approach focuses on piecemeal improvements of the model while keeping the data fixed. The intuition is that so long as a model is faithfully reproducing target system behavior, refining the model will produce increasingly improved fit with the target system's behavior. Target systems and model trajectories are expected to converge monotonically as the models are made more realistic.

Both approaches share piecemeal monotonic convergence of model output to target system behavior under the faithful model framework. The intuitive appeal of such piecemeal strategies is clear for linear models: any small change in the magnitude of an observable is guaranteed to yield a proportionately small change in the output of the model. For faithful linear models, making small improvements 'in the right direction' in either data or model accuracy can be tracked by improved model performance in state space. The qualifier 'in the right direction' draws directly on the faithful model framework and means that the data quality really is increased or that the model really is more realistic—captures more features of the target system in an increasingly accurate way.

Nonlinear models contrast sharply with this intuitive picture. When SDIC becomes an issue, piecemeal approaches prove problematic. Suppose we improve the initial data to some degree. Such improvement isn't guaranteed to produce more convergence between imperfect model behavior and target system behavior. Sensitive dependence implies small refinements in data quality in the right direction

---

[7] For an early discussion in weather forecasting, see Thompson [7].

aren't guaranteed to lead to a nonlinear model monotonically improving in describing target system behavior. Similarly, making successive refinements in nonlinear models, while keeping data fixed, isn't guaranteed to lead to more convergence between model output and target system behavior.[8]

Note that there are no violations of determinism; nonlinear dynamical systems can exhibit unique evolution while still suffering from the effects of the loss of linear superposition. Sensitive dependence renders model construction, refinement, use, and interpretation more challenging than contexts involving linear dynamical systems. Scientists account for uncertainty in data and model inadequacy in nonlinear contexts as we'll see.

There is a further issue with the faithful model framework: what is the relationship between the mathematical model and the actual-world target system? Typically, scientists assume a one-to-one relationship between model and target system. The implication is that observables and parameters and how they change with time in the model equations can be directly related to actual features of the target system and vice versa. However, it may be the case that the actual relationship is one-to-many, with several different nonlinear models describing the same target system or, potentially, several different target systems being described by the same nonlinear model. Or perhaps the relationship between models and actual-world systems is a many-to-many relationship.

For linear systems, such as a mass on a spring executing small oscillations, the relationship or translation between model and target system appears to be straightforwardly one-to-one. In contrast, consider nonlinear contexts, where one constructs a model from a time series generated by observing a system. There are potentially many nonlinear models that can be constructed, each as empirically adequate to the system behavior as any other. Is there only one unique model for each target system and we simply do not know which is the true one? Are there no one-to-one relationships between our mathematical models and target systems? Scientists and philosophers of science don't know the answer to these questions. This is another challenge scientists seeking to model actual-world phenomena face.

1.9.1 The governance myth

As a final remark on modeling, it's not unusual to read scientists talking about how equations, models, and laws 'govern' target system behaviors. This is the **governance myth**, which makes it sound like target systems must obey the equations we use to describe them. Nevertheless, equations, models, and laws don't govern or control anything.[9] Consider an analogy: the number π characterizes the geometry of circles and spheres but doesn't control what frisbees and balls are or how they behave. It represents a constraint on or an affordance for their shapes. Similarly, the universality of Feigenbaum constants (section 3.1.2) across so many mathematical one-hump maps and disparate physical systems doesn't imply it possesses any governing

---

[8] I have witnessed this in plasma physics modeling.

[9] For more discussion and examples where the governance metaphor fails, see Bishop [8].

control over these systems. Such universality indicates that a wide variety of systems share some key structural features.

We can never prove that our equations, models, or laws govern anything. Rather, we can only demonstrate that they describe actual-world phenomena well. Consider a simple example: suppose a hunk of iron is dropped from a height. It will fall as Newton's gravitational force law describes in the absence of any other relevant factors. Suppose further that during its fall an appropriately configured electromagnetic field is switched on. Now the hunk remains suspended in space rather than falling to the ground. Newton's gravitational force law isn't 'governing' anything. But neither is the electromagnetic field governing anything. The iron hunk's behavior is adequately described by the combination of Newton's and Maxwell's laws. Now if I reach out and grab the hunk with my hand, neither Newton's nor Maxwell's laws describe the behavior of the iron.

With these preliminaries in place, we can now turn to the history and phenomenology of chaotic dynamics.

References

[1] Duhem P 1906 *The Aim and Structure of Physical Theory* (Princeton, NJ: Princeton University Press) https://doi.org/10.1515/9780691233857
[2] Bishop R C 2003 On separating prediction from determinism *Erkenntnis* **58** 169–88
[3] Bishop R C 2024 *The Stanford Encyclopedia of Philosophy* (Winter 2024 Edition), E N Zalta and U Nodelman (eds.), https://plato.stanford.edu/archives/win2024/entries/chaos/
[4] Lorenz E N 1963 Deterministic nonperiodic flow *J. Atmos. Sci.* **20** 130–41
[5] Poincaré H 1908 *The Foundations of Science: Science and Method* **1913** (Lancaster: The Science Press)
[6] Rössler O 1983 The chaotic hierarchy *Z. Naturf. A* **38** 788–801
[7] Thompson P D 1957 Uncertainty of initial state as a factor in the predictability of large scale atmospheric flow patterns *Tellus* **9** 275–95
[8] Bishop R C 2024 *The Physics of Emergence* 2nd ed (Bristol: IOP Publishing) https://doi.org/10.1088/978-0-7503-6367-9

Chapter 2

A brief history of chaotic dynamics

This chapter introduces some of the history leading up to the formation of the field of chaos studies enabling us to see this field of study in wider context. The main focus is on the rise of interest in sensitive dependence and aperiodicity.

One of the earliest appearances of the term 'chaos' in the physics literature is in a paper on dynamical systems by Bernard Koopman (1900–81) and John von Neumann (1903–57) [1]. Although in the context of the behavior of trajectories spreading densely throughout state space, their work didn't address the modern phenomenon of chaotic dynamics.

2.1 Sensitive dependence on initial conditions (SDIC)

Aristotle (384–22 BCE) understood that small changes could lead to larger effects in time. Writing about methodology and epistemology in mathematics, he observed that 'the least initial deviation from the truth is multiplied later a thousandfold' [2], 271b8. While it's true that small deviations can lead to significantly different outcomes in mathematics, arguably Aristotle doesn't articulate enough constraints for us to identify this as an expression of the sensitive dependence on initial conditions (SDIC) associated with chaos.

Prior to the 20th century, few physicists and mathematicians explored the phenomenon of small initial changes leading to significantly different behavior in systems. Such investigations were isolated and never produced a sustained field of inquiry. For example, James Clerk Maxwell (1831–79) identified behavior resembling SDIC [3], p 13. He described such phenomena as cases where the 'physical axiom' that from like antecedents flow like consequences is violated. Maxwell recognized this kind of behavior could be found in systems with a sufficiently large number of variables. But he also argued that such sensitive dependence could happen in the simple case of two colliding spheres [4].

Poincaré recognized that this same kind of behavior could be realized in systems with a small number of variables (simple systems exhibiting very complicated behavior). In his discussion of chance in *Science and Method*, he notes,

doi:10.1088/978-0-7503-6453-9ch2

> But even when the natural laws should have no further secret for us, we could know the initial situation only *approximately*. If that permits us to foresee the subsequent situation *with the same degree of approximation*, this is all we require, we say the phenomenon has been predicted, that it is ruled by laws. But this is not always the case; it may happen that slight differences in the initial conditions produce very great differences in the final phenomena; a slight error in the former would make an enormous error in the latter [5], pp 397–8; emphasis original.

This is the idea that even the smallest uncertainty in our measurement of the initial state of a system can lead to a very large uncertainty in a later or final state. This approaches how some definitions of chaos work (section 5.2).

Yet, in hindsight, Poincaré discussed examples that can be viewed as raising doubts about explosive growth in uncertainty as a sufficient condition for characterizing chaos. He discussed the case of a perfectly symmetric cone precisely balanced on its tip with only the force of gravity acting on it. As we've seen, the sensitivity in such cases doesn't yield SDIC (section 1.8). Variations in the slightest causes issue forth in dramatically different effects (a violation of the physical axiom Maxwell articulated), and fits the way Poincaré described the growth in uncertainty in the quote given above. Nonetheless, this behavior isn't the same, as the SDIC mathematicians and scientists describe.[1] His cyclone example (section 1.8) is much closer.[2]

Duhem, relying on work by Poincaré as well as Jacques Hadamard (1865–1963), further articulated the practical consequences of sensitive dependence for the scientist interested in deducing mathematically precise consequences from mathematical models [8], p 138–42. Hadamard had developed the framework for partial differential equations exhibiting both continuous and discontinuous dependence on initial conditions by 1922. Any equations exhibiting sensitive but continuous dependence are well-posed problems under his framework; however, he raised the possibility that any solution to equations for a physical system exhibiting such sensitive dependence could indicate that the target system obeyed no laws [9], p 38.

Nevertheless, the early work by mathematicians and physicists on such sensitivity remained mostly a mathematical curiosity, attracting little sustained attention. It remained unrecognized that such behavior needed systematic treatment, though there were tantalizing hints that were never pursued at the time (e.g., [10–12]).

2.1.1 Edward Lorenz and the 'discovery' of chaos

Since weather forecasts have important economic and health consequences, significant resources have been poured into developing accurate weather forecasting

---

[1] Movie goers may recognize this example as being similar to that given by mathematician Ian Malcolm in the movie Jurassic Park to explain chaos, though his discussion in the movie confuses chaos with a different kind of sensitivity [6].

[2] As is Poincaré's discussion of the stability and predictability of Solar System dynamics [7].

models for over a century.[3] The interest in improving the accuracy of forecasting models has been intense and Edward Lorenz (1917–2008) decided to put the debates over different approaches to modeling to numerical test. He aimed to test two main approaches for weather forecasting, one involving linear modeling and the other nonlinear modeling.

Competing forecasting approaches couldn't be distinguished using periodic solutions, so he developed simplified weather forecasting models to search for aperiodic solutions (section 1.6). After exploring a 12-equation model, with the help of a colleague Lorenz tried the simpler set of three equations, the Lorenz model (section 1.7) forever identified with his name.

In 1960, Lorenz's model ran on a computer the size of a desk. The computer was so loud it had to be put in its own office. The actual computer programs were written and run by Margaret Hamilton and Ellen Fetter. This task required creating the binary code on long spools of paper tape [13], a feat of mathematical and computer science skill crucial to the birth of chaos studies. After some trial and error, Lorenz found parameter settings for the model that produced aperiodic solutions. He demonstrated that the results of the nonlinear modeling approach could not be reproduced by linear approaches.

Yet, something surprising happened when Lorenz decided to rerun the model with some of the model's own output from an earlier time to dive into more detail about how the weather in the model was behaving. This was equivalent to backing the model weather up several weeks to run again, but the computer output diverged from the original run very rapidly.

The culprit was in the data used to rerun the model. Originally, the computer had been keeping track of data to six decimal places, but the printout contained data rounded to the third decimal place to save paper. When printout data was used to re-initialize the computer, the small difference in rounding the input had been amplified rapidly by the computer model.[4] Lorenz had found chaos, though he didn't give it that name. The earliest use of the term 'chaotic' to describe the phenomenon Lorenz observed was in 1971 in David Ruelle and Floris Takens' (1940–2010) [16].[5] It was an influential 1975 paper by Tien-Yien Li (1945–2020) and James A Yorke [17] that led to the widespread use of the term 'chaos' for these mathematical behaviors.

To illustrate the phenomenon, Lorenz adopted a simpler set of model equations that captured some features of fluid convection in the atmosphere leading to his

---

[3] Weather models were among the first applications for the first computers.

[4] His famous illustration of a butterfly in Brazil causing a tornado in Texas comes from a 1972 American Association for the Advancement of Science talk Lorenz gave [14], appendix 1, where he described the rapid growth of errors in the initial conditions for cyclonic weather systems. A possible allusion to a 'grasshopper effect' was made by W S Franklin in an 1898 review of Duhem's *Traité Élémtaire de Méchanique Chimique fondée sur la Thermodynamique*. Franklin believed long-range weather forecasting would be impossible because weather prediction 'is subject to the condition that the flight of a grasshopper in Montana may turn a storm aside from Philadelphia to New York!' [15], p 173, though he never specified any growth rate for the uncertainty introduced by a grasshopper. As we'll see in chapter 5, such ambiguity makes it difficult to infer chaotic dynamics from Franklin's example.

[5] The term strange attractor also first appears in this paper.

path-breaking [18]. This publication was a foundational paper kicking off the study of chaotic dynamics. While there is a case to be made that Lorenz 'discovered' mathematical chaos, as can be seen from Poincare's remarks above, some level of awareness of the phenomenon had been around for decades. Mathematicians, such as Aleksandr Lyapunov (1857–1918), Philip Franklin (1898–1965), and Andrey A. Markov (1856–1922) also had discussed conditions for stability and proved when solutions to differential and partial differential equations were stable. The implications for chaotic dynamics were there when solutions failed to meet these stability conditions. Nonetheless, the implications remained largely unrecognized and unexplored.

2.2 Aperiodicity

Periodic solutions are attractors repeating an identical pattern over and over with regularity: $x(t)$ is periodic if $x(t + T) = x(t)$ for some $T > 0$. In contrast, aperiodic solutions don't repeat their behavior. Although aperiodicity is an important feature of chaotic dynamics, such behavior was not considered to be of systematic interest until the 1960s.

The norm in 19th and 20th century mathematics and physics was to focus on periodic solutions to equations and ignore aperiodic behavior as transitory. This approach is illustrated by Hadamard's [9], which never mentions aperiodic solutions. In a 1927 article, George Birkhoff (1844–1944) [19] discusses almost periodic orbits, solutions converging to periodic in an appropriate limit. In this same article, he notes conditions where no periodic orbits exist (e.g., p 376), but doesn't consider them of interest. Birkhoff demonstrates that in cases where there are periodic orbits near aperiodic ones, under appropriate conditions periodic orbits always outnumber aperiodic orbits.[6]

Philip Franklin proved in 1929 that solutions originating sufficiently close remains close over time. The implication is that such solutions must be periodic or almost periodic. In contrast, Franklin's proof implies that solutions failing to have this property are aperiodic and must be subject to sensitive dependence, Franklin didn't note this implication of aperiodic solutions, however [10].

In 1932, Koopman and von Neumann [1] observed that for dynamical systems with self-adjoint unitary operators possessing a continuous spectrum of eigenvalues, where $U_t f = f$ almost everywhere in some region Ω in state space S, then if there is no invariant subset of points in Ω of finite measure, and Ω itself has finite measure, trajectories initiating from any neighborhood $N \in \Omega$ will spread out well beyond N. The upshot is that 'Periodic orbits, and such like, appear only as very special possibilities of negligible possibility' (p 261). Such behavior could be exhibited in systems with as few as two degrees of freedom, so is distinct from the behavior in statistical mechanics systems of many degrees of freedom.

In 1945, Dame Mary Lucy Cartwright (1900–1998) and John Littlewood (1885–1970) [20] also noted unstable periodic orbits as well as aperiodic orbits can be a general feature of nonlinear equations depending on the parameter magnitudes of key nonlinear terms, using

---

[6] In fact there are infinitely many more periodic orbits for each aperiodic orbit.

$$\ddot{y} - k(1 - y^2)\dot{y} + y = b\lambda k \cos(\lambda t + a) \qquad (2.1)$$

as their example. Interestingly, in the final section of their article they mention the spreading of unstable periodic trajectories issuing forth from nearby initial conditions, but don't explore the rate of this spreading. The rate of uncertainty growth has been an important indicator, even definition of, chaos (chapter 5). Nevertheless, Cartwright and Littlewood's work helped popularize the applications of Poincaré's methods to dynamical systems and provided important background for Lorenz's studies.

As with SDIC, there was work revealing aperiodicity for dynamical systems under various conditions, but the phenomenon was treated as an uninteresting mathematical oddity at best. The exceptions were circuits and radio engineering, which involved nonlinear differential equations, and dynamical system with an aperiodic forcing function. An oscillator, such as a pendulum, will execute aperiodic motion corresponding to the aperiodic external forcing. But this fact is not surprising.

The situation changed with the publication of Lorenz's [18]. Normally, one cannot demonstrate that numerical solutions are absolutely aperiodic because it's not possible to prove there is no recurrence time (the time when a solution will repeat or nearly repeat itself). Lorenz was able to infer that the aperiodic solutions of his model wouldn't repeat themselves because they had a deep relationship to the Cantor set [21].[7] This inference has been confirmed [22].

With Lorenz's landmark paper, aperiodic solutions began to be studied systematically and, like SDIC, became indelibly associated with chaotic dynamics. For instance, Physicist Boris Chirikov (1928–2008) investigated a two-dimensional dynamical system that exhibited aperiodic solutions under various parameter settings (though he never references Lorenz's work) [23] (section 3.2).

The stretching and folding dynamics mentioned in section 1.8 as important for chaos also is related to aperiodicity. Orbits that don't repeat themselves have a tendency to fill the state space over time (this is the effect of the stretching of trajectories away from each other). But, as noted earlier, these orbits don't leave state space because of the folding processes. The stretching and folding of aperiodic orbits ensures both that trajectories are bounded in some region of state space while also never repeating themselves nor intersect themselves, otherwise system observables would become multiply valued and have multiple future paths, a violation of unique evolution (section 1.3).

References

[1] Koopman J and von Neumann J 1932 Dynamical systems of continuous spectra *Proc. Natl Acad. Sci.* **16** 255–61

[2] 1985 *The Complete Works of Aristotle: The Revised Oxford Translation* **vol 1** ed J Barnes (Princeton, NJ: Princeton Univ. Press) https://doi.org/10.1515/9781400835843

---

[7] In essence, Lorenz inferred that the attractor for the aperiodic behavior of his solutions was a strange attractor.

[3] Matter M J C 1992 *Matter and Motion* (New York: Dover) 1876

[4] Maxwell J C 1965 Illustrations of the dynamical theory of gases *The Scientific Papers of James Clerk Maxwell* (New York: Dover) 378–9 pp

[5] Poincaré H 1908 *The Foundations of Science: Science and Method* (Lancaster: Science Press) 1913

[6] Carpineti A Why was Ian Malcom, a mathematician, invited to Jurassic Park? *IFLScience* 14 June 2023. https://www.iflscience.com/why-was-ian-malcolm-a-mathematician-invited-to-jurassic-park-69373.

[7] Poincaré H 1993 New methods of celestial mechanics *History of Modern Physics and Astronomy* (Melville, NY: AIP)

[8] Duhem P 1906 *The Aim and Structure of Physical Theory* (Princeton, NJ: Princeton Univ. Press) 1982

[9] Hadamard J 1922 *Lectures on Cauch's Problem in Linear Partial Differential equations* (New Haven, CT: Yale Univ. Press)

[10] Franklin P 1929 Almost periodic recurrent motions *Math. Z.* **30** 325–31

[11] Nemytskii V V and Stepanov V V 1960 *The Quantitative Theory of Differential equations* (Princeton, NJ: Princeton Univ. Press)

[12] Hénon M and Heiles C 1964 The applicability of the third integral of motion: some numerical experiments *Astron. J.* **69** 73–9

[13] Sokol J 2019 The hidden heroines of chaos *Quantamagazine* https://www.quantamagazine.org/the-hidden-heroines-of-chaos-20190520.

[14] Lorenz E N 1995 *The Essence of Chaos* (Seattle, WA: Univ. Washington Press)

[15] Franklin W S 1898 New books *Phys. Rev.* **6** 170–5

[16] Ruelle D and Takens F 1971 On the nature of turbulence *Commun. Math. Phys.* **20** 167–92

[17] Li T-Y and Yorke J A 1975 Period three implies chaos *Am. Math. Mon.* **82** 985–92

[18] Lorenz E N 1963 Deterministic nonperiodic flow *J. Atmos. Sci.* **20** 130–41

[19] Birkhof G 1927 On the periodic motions of dynamical systems *Acta Math.* **50** 359–79

[20] Cartwright M L and Littlewood J E 1945 On non-linear differential equations of the second order: I. The equation $\ddot{y} - k(1 - y^2)\dot{y} + y = b\lambda k \cos(\lambda t + a)$, k large *J. London Math. Soc.* **20** 180–9

[21] Vallin R W 2013 *The Elements of Cantor Sets: With Applications* (Hoboken, NJ: Wiley) https://doi.org/10.1002/9781118548745

[22] Viswanath D 2004 The fractal property of the Lorenz attractor *Physica* D **190** 115–28

[23] Chirikov C V 1969 *Research concerning the theory of nonlinear resonance and stochasticity.* CERN-Trans-71-40 Institute of Nuclear Physics, Novosibirsk http://cds.cern.ch/record/325497?ln=en

Chapter 3

The phenomenology of chaotic dynamics

With the conceptual and historical background in place, this chapter illustrates the phenomenology of chaotic behaviour using model conservative dynamical systems (section 1.6), such as the logistic and Chirikov standard maps. Chaos is an intricate behavior found in such simple systems. Although observables characterizing these systems are important, the role of parameters for the emergence of chaotic behavior is equally important.

3.1 A simple example: the logistic map

In 1976, physicist turned ecologist Robert M May (1936–2020) published a seminal review article in *Nature* [1], collecting in one place the variety of complexity and order exhibited by mathematical chaos. His review focused on a simple ecological example. Consider a seasonal breeding population, where the generations don't overlap. For instance, a crop pest may produce a new generation every season that a crop is growing but doesn't breed year-round. One might want to predict the size of this year's population based on the size of last year's. This relationship turns out to be nonlinear (section 1.7). Moreover, this year's population size may be directly related to last year's in a relationship describable by a deterministic model (section 1.3).

For a mathematical model with these characteristics, if a previous generation's population is small, population size increases monotonically in subsequent generations. When a previous generation's population is too large, population size decreases monotonically in subsequent generations. Under many conditions, then, the model for population growth either increases strictly monotonically or decreases strictly monotonically generation by generation.

The simplest mathematical model capturing this behavior is the following: let x_t represent the population of the current generation and t represent this year. The size of the next year's population is related to the size of this year's population by the following rule, called the **logistic map**:

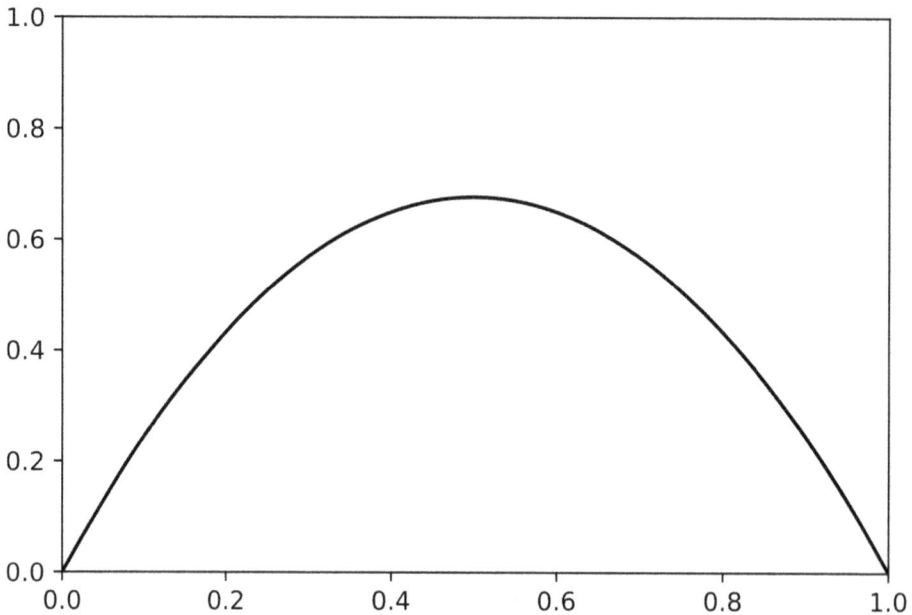

Figure 3.1. The logistic map, plotted for $\alpha = 2.707$, rises to a maximum at $x = 0.5$.

$$x_{t+1} = \alpha x_t (1 - x_t), \qquad (3.1)$$

where α is a parameter ranging in value from one to four representing the density of resources available to the population. Figure 3.1 illustrates the behavior of the logistic map where a small x_t leads to a larger population in the next generation x_{t+1}, while a population x_t that is too large will produce a smaller generation at x_{t+1}. Equation (3.1) is an example of a nonlinear dynamical system. It is nonlinear because of the presence of x^2 (anytime observables are multiplied by observables an equation will be nonlinear even if it is one-dimensional). Although a simple mathematical model, its behavior turns out to be surprisingly complex and rich.

The logistic map has an attractor for all magnitudes of $\alpha < 4.0$. One can see this from looking at figure 3.1: the maximum value x can take is $\alpha/4$ since x is confined to the range [0, 1].

Therefore, this map has an attractor under this condition. For small magnitudes of α, the attractor is $x = 0$. For any $\alpha > 1$, magnitudes of $x > 0$ all move away from zero: the characteristics of the attractor depend on the magnitude of α. This makes sense for a population model. When a population is small, it underuses resources leaving more for the next generation, but too large a population overuses resources leaving less for the next generation.

3.1.1 Stability

It is easy to see that when $\alpha = 0$ and the initial value $x_o = 0.5$, the logistic map has a fixed-point attractor (section 1.6). The uncertainty in the initial conditions in this case will cease growing as the trajectories in state space converge to zero. How

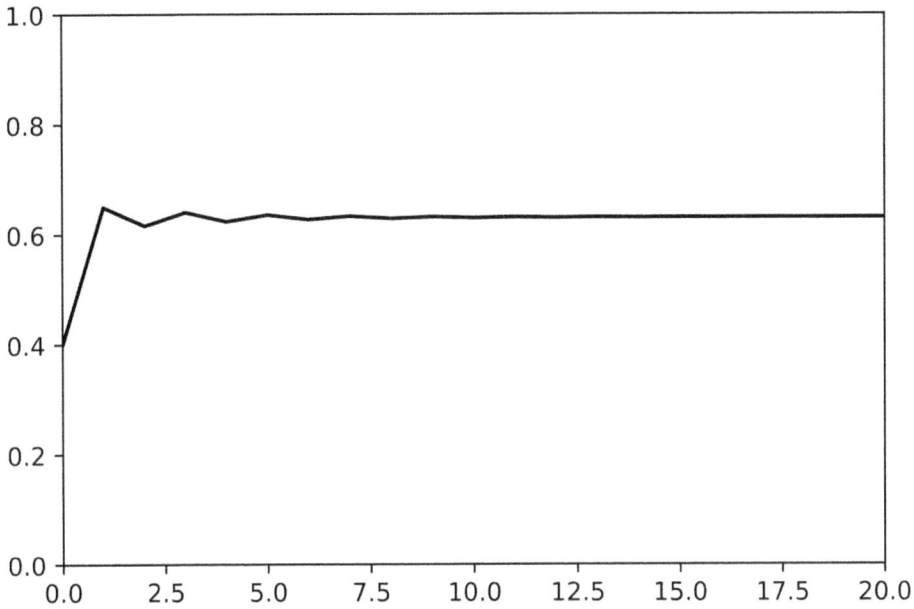

Figure 3.2. Stability of the logistic map for $\alpha = 2.707$. The map settles into a point attractor after a few iterations.

stable is this attractor? The fixed-point attractor is stable for all magnitudes of $0 \leqslant \alpha < 3.0$. For instance, compare figures 3.1 and 3.2.

What happens as α approaches three? Suppose $\alpha = 2.9$. Comparing figures 3.2 and 3.3, observe that the logistic map takes much longer to converge to the point attractor as α approaches three. At $\alpha = 3$, a period doubling **bifurcation** occurs, where the logistic map settles into a period loop attractor of period two, sometimes known as a pitchfork bifurcation since the system can now take two paths in state space. A bifurcation is a qualitative change in the behavior of a dynamical system as a parameter, such as α, changes smoothly.

Consider $\alpha = 3.1$, for instance (Figure 3.4). The fixed-point attractor hasn't vanished; rather, it's now an unstable fixed point. For $\alpha > 3$, any trajectory approaching the fixed-point attractor will not settle down there but follow one branch or the other of the bifurcation.

As α increases to 3.5, another bifurcation occurs, where the map settles into a period four attractor. Further increases to α lead to successive bifurcations— attractors of period eight, then 16, then 32, then 64, and so forth. In other words, the attractor period doubles with every bifurcation. Figure 3.5 illustrates several of these period doublings: periods 2, 4, 8, and 16 for corresponding magnitudes of increased α. As with the fixed-point attractor, the period two and successive attractors don't vanish; they each become unstable in turn and trajectories approaching them will follow one or another branching path of the new loop attractor.

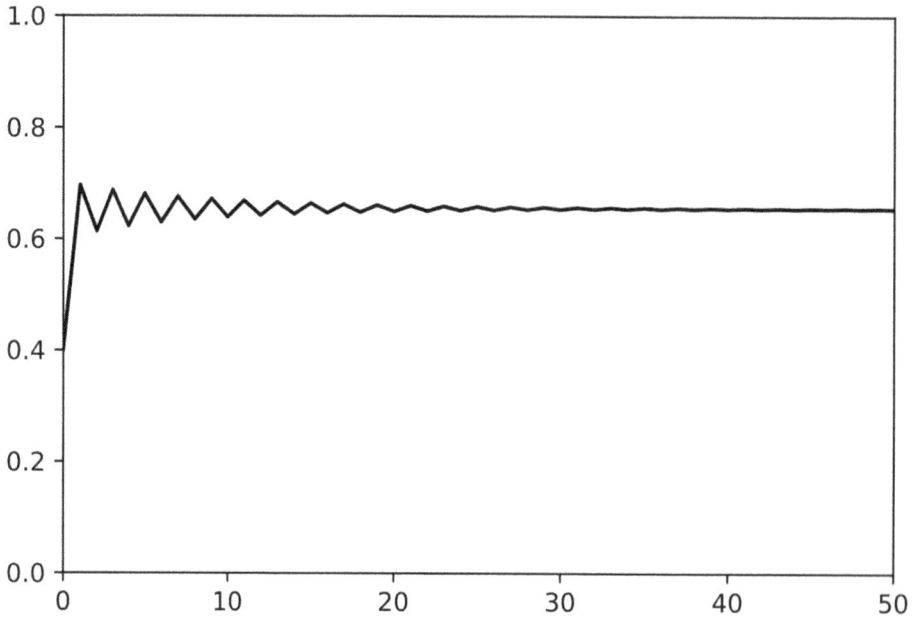

Figure 3.3. Stability of the logistic map for $\alpha = 2.9$. The logistic map takes much longer to settle down to a point attractor as α approaches three.

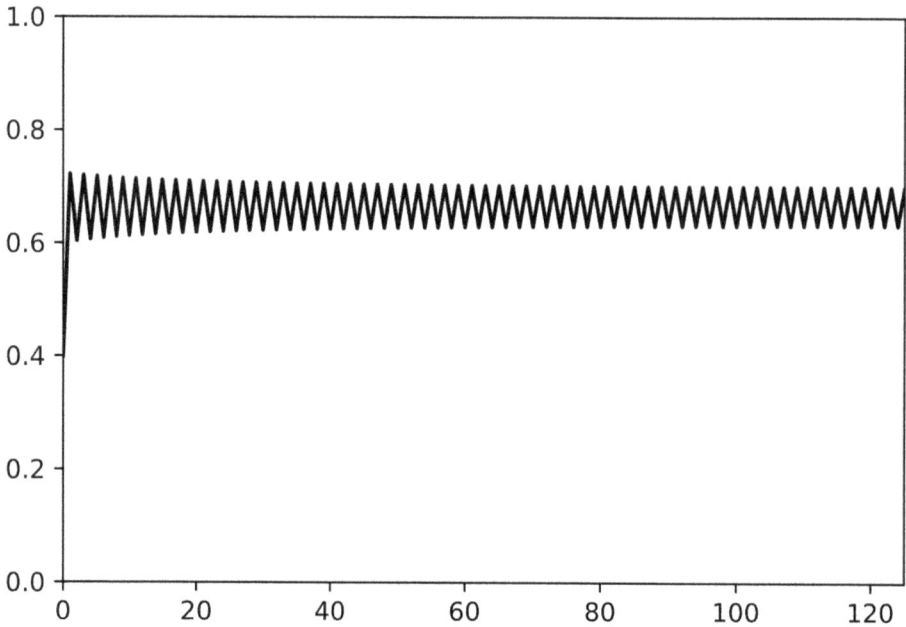

Figure 3.4. At $\alpha = 3.01$, the logistic map's bifurcates to a period two attractor.

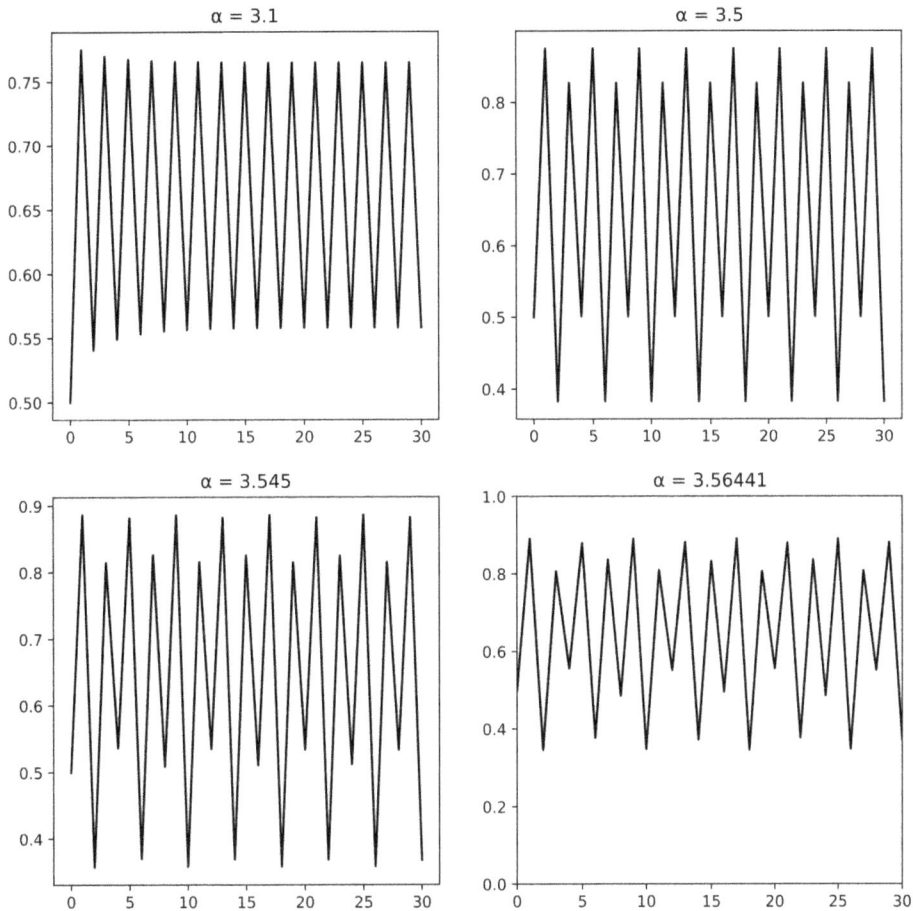

Figure 3.5. A sequence of logistic map plots illustrating period doubling bifurcations for increasing α: period two ($\alpha = 3.1$), period four ($\alpha = 3.5$), period eight ($\alpha = 3.545$), and period 16 ($\alpha = 3.564\,41$).

3.1.2 Chaos

Comparing the results in figure 3.5, the magnitudes of α get closer together for each successive bifurcation. As α increases to 3.7 (Figure 3.6), outputs for successive iterations no longer repeat with any period that we can discern—there are no attractors. In other words, the logistic map now yields aperiodic outputs. This is the realm of chaotic dynamics. Even though the behavior appears random, it's still deterministic (apparent randomness, section 1.1) because it's generated by a deterministic equation.

Moreover, For $\alpha = 3.7$, the logistic map exhibits sensitive dependence on initial conditions (SDIC) (section 1.8). Figure 3.7 illustrates the difference in the map's behavior when the initial condition is changed by one ten thousandth. Map iterations behave the same for a while, but the trajectories begin to diverge from each other at about time step 50. Trying this same exercise with any of the magnitudes of $\alpha < 3.7$ yields no difference in trajectories whatsoever. We have SDIC, one of the properties characterizing chaos.

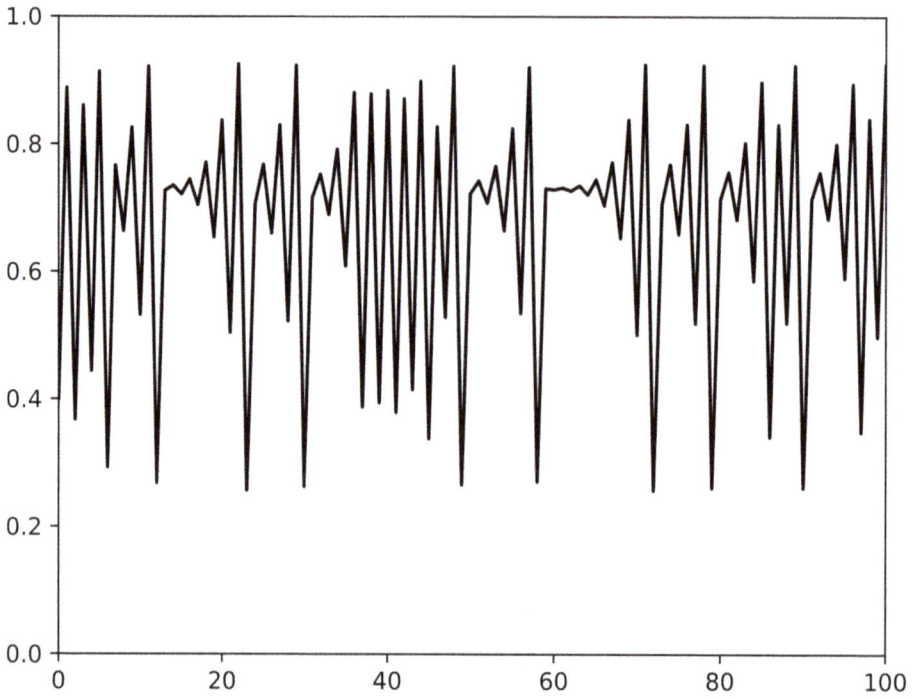

Figure 3.6. For $\alpha = 3.7$, the logistic map exhibits aperiodic behavior.

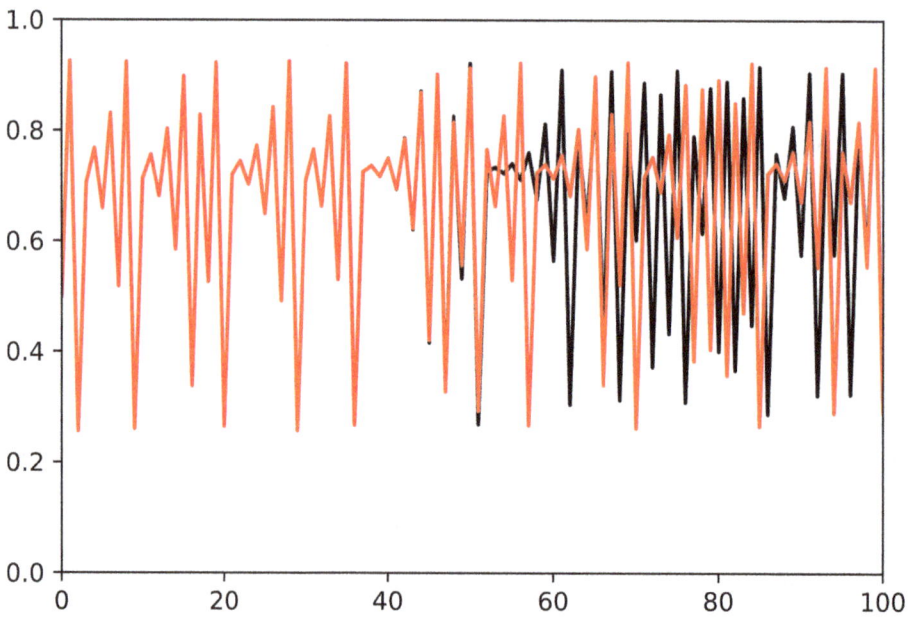

Figure 3.7. The logistic map for $\alpha = 3.7$ with two initial conditions differing by one ten thousandth.

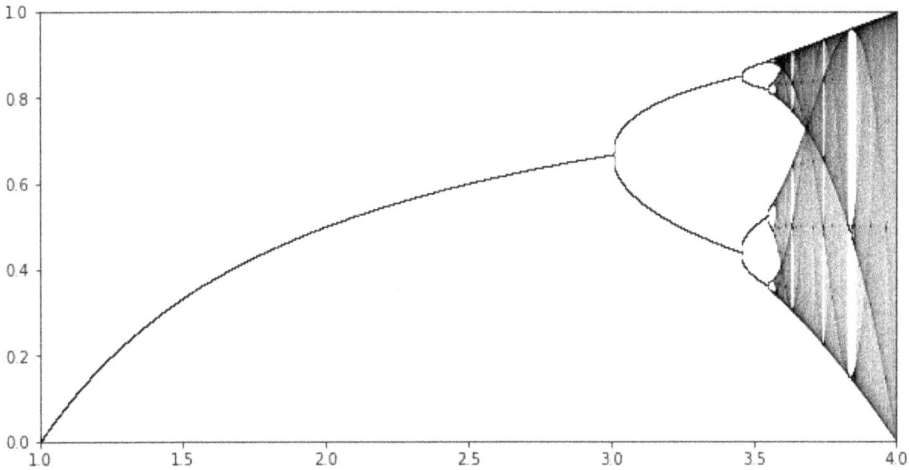

Figure 3.8. Bifurcation diagram showing the sequence of period doublings for the logistic map as α increases leading to chaos.

Clearly the magnitude of the parameter α conditions when the logistic map exhibits chaotic dynamics. A perspicuous way of seeing this is by plotting the bifurcation points in a **bifurcation diagram** (Figure 3.8), a plot of fixed or periodic points versus a parameter.

The output of the logistic map is plotted versus α. For $0 < \alpha < 3$, the map exhibits a fixed-point attractor. At $\alpha = 3$, period doubling occurs. As α increases, period doubling events leading to a period 4 attractor, then period 8, then period 16 and so forth can be observed.

As α increases, the period doublings come increasingly rapidly and eventually transition to period three attractors starting at period three and then bifurcating to six, then to 12 and so on. These period three bifurcations likewise become increasingly rapid as α increases until period four bifurcations start, then period five and so on through all the integers until the plot looks like a jumbled mess. This 'jumble' is the chaotic behavior of the logistic map revealing an exquisite sensitive dependence on the parameter α.

While the right-hand side of figure 3.8 looks random, it is filled with complicated order as figure 3.9 reveals.

Focusing on the bifurcation diagram for $3.7 < \alpha < 3.9$, there are regions of apparently random behavior in the logistic map. These are home to aperiodic trajectories, where the trajectories never repeat their behavior (so far as we can determine computationally). Nonetheless, this behavior is punctuated by regions where period loop attractors reemerge of every integer period. For example, period doubling sequences emerge once again, and at $\alpha = 3.83$ a period three orbit emerges.

There is another regularity hidden from view in figures 3.8 and 3.9. In 1975, Mitchell Feigenbaum [2] derived a surprising result examining the period doubling

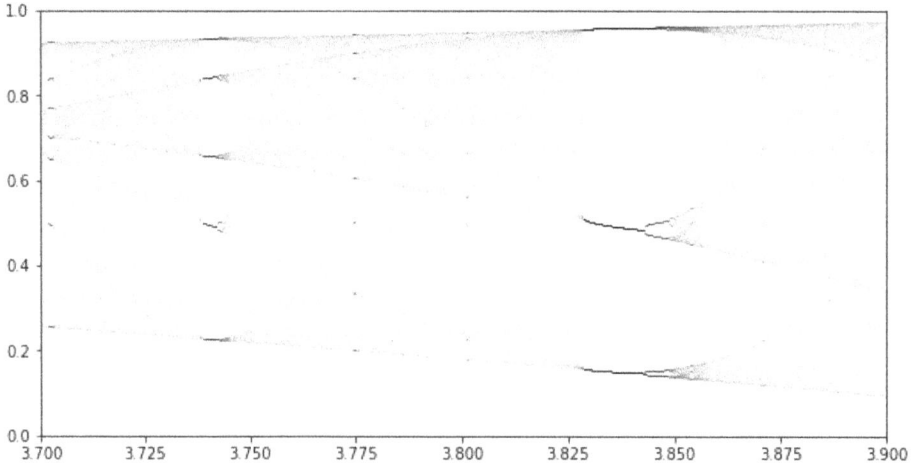

Figure 3.9. Zooming into the bifurcation diagram for magnitudes of $3.7 < \alpha < 3.9$ reveals intricate order characteristic of chaotic dynamics.

Table 3.1. Feigenbaum constant for successive values of α computed for the period doubling sequence for the logistic map.

Period	α	δ
2	3.0	N/A
4	3.449 490	N/A
8	3.544 090	4.7514
16	3.564 407	4.6562
32	3.568 759	4.6683
64	3.569 692	4.6686
128	3.569 891	4.6692
256	3.569 934	4.6694

sequence for the logistic map. In the limit as $n \to \infty$, the ratio of each bifurcation window to the next window between each period doubling is a constant:

$$\delta \equiv \lim_{n \to \infty} \frac{\alpha_{n+1} - \alpha_n}{\alpha_{n+2} - \alpha_{n+1}} \tag{3.2}$$

where $\delta = 4.669\,201\,6091\ldots$. This result actually holds for any map with a single local quadratic maximum. An implication of this result is that the parameter magnitudes α for which period doubling bifurcations occur get closer together as table 3.1 illustrates for the first few values of α.

Another feature of chaos' intricate behavior is that figures 3.8 and 3.9 suggest there are many parameter magnitudes for which the logistic map behaves chaotically. Indeed, there are an infinite number. Nonetheless, compared to the total number of parameter magnitudes for which there are stable attractors, the number of magnitudes yielding chaotic behavior is of measure zero. This roughly means that while infinitely many magnitudes of α exist leading to chaotic behavior, there is a much larger infinity leading to stable attractors. Although true for the logistic map, this result is generally true for all chaotic maps and flows: there are far more parameter magnitudes leading to stable behavior than chaotic. Hence, any claim that chaotic behavior is 'ubiquitous' is misleading without significant nuance.

The complex order characteristic of chaotic dynamics is far from lawless randomness (section 1.1). Complicated chaotic behavior can arise in a very simple mathematical model. As Maxwell and Poincaré noted (section 2.1), complicated models are not needed to study complex behavior. Moreover, note that the logistic map's chaotic dynamics exhibit sensitivity to the magnitudes of x and α. The magnitude of α determines when SDIC appears. When it does, the smallest changes in x yields dramatic changes in the behaviors of the system (e.g., Figure 3.7). Such sensitivity to both observable and parameter magnitudes is a general feature of models exhibiting chaotic behavior.

3.2 A more complex example

The Chirikov standard map[1] [3, 4] is a two-dimensional dynamical system, but, like the logistic map, is conservative or state space area preserving:

$$P_{t+1} = P_t + \sin \theta_t, \tag{3.3}$$

$$\theta_{t+1} = \theta_t + P_{t+1}, \tag{3.4}$$

where P and θ are both computed mod (2π) so that the mapping always stays on the unit square. This is a model for a periodically kicked rotor fixed at one end and rotating friction free. There is no gravity in the system, P represents the rotor's momentum, while θ represents its angular position, and K is the magnitude of the external kick. Equations (3.3) and (3.4) describe how the momentum and angular position of the rotor at time $t + 1$ depend on their magnitudes at t plus the effect of the external forcing.

Although a simple model, elementary particles orbiting in circular accelerators are well described by the Chirikov map. Protons, say, are periodically kicked by an electromagnetic field accelerating them around the tunnel, while friction and gravity effects are negligible. As with the logistic map, the Chirikov map is a simplified but reasonable mathematical model for gaining insight into a physical situation.

The Chirikov map has two fixed-point attractors when $P_{t+1} = P_t$ and $\theta_{t+1} = \theta_t$, namely when $P = 0$ and $\theta = 0$ or π. The fixed-point attractors all are stable for $K = 0$ (no kick disturbing the rotor). These conditions correspond to periodic orbits.

---

[1] For historical reasons, the term 'map' is used of this two-dimensional system even though it technically is a flow. This dynamical system also is called the Chirikov–Taylor standard map as it was independently discovered by Bryan Taylor (1928–2004).

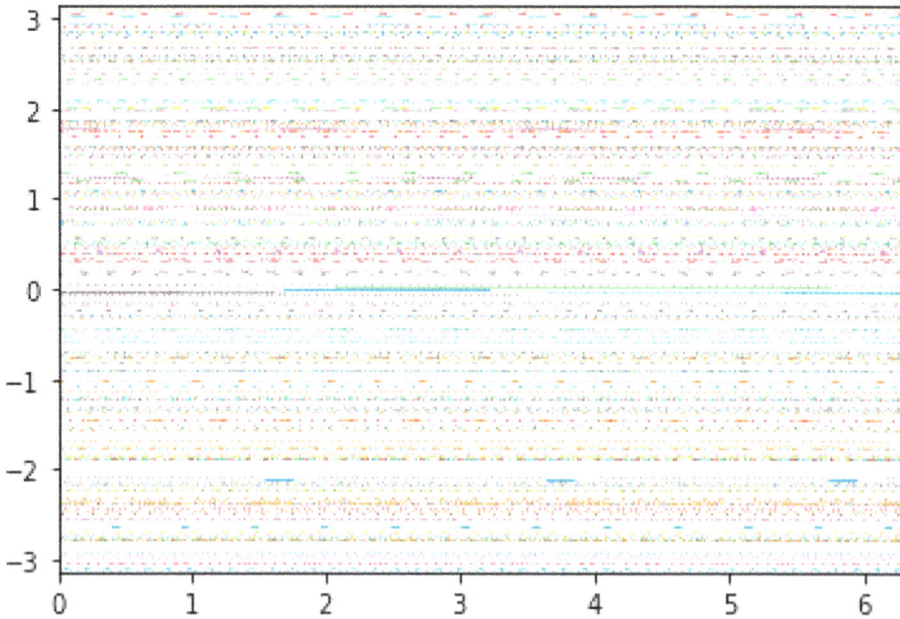

Figure 3.10. The Chirikov map plotting θ versus P for $K = 0$ displaying periodic and quasi-periodic orbits.

Plotting a variety of initial conditions for $K = 0$ gives a feel for the map's behavior in the absence of external forcing (Figure 3.10).

This figure is known as a **Poincaré surface of section** plot: slice a perpendicular plane through the trajectories so that each dot in the figure represents a trajectory coming out of the page at you (i.e., particles are moving in a closed loop as you observe them head on). More formally, a Poincaré surface of section plot for a d-dimensional system is generated by the flow—the orbits—intersecting with a $d-1$ dimensional submanifold. This produces what is often called a **return map**.

For $K = 0$, the Chirikov map is linear. Depending on the initial conditions, it produces either periodic or quasi-periodic orbits. The periodic orbits are closed loops and show up in figure 3.10 as equally spaced dots, where each dot represents a particle trajectory repeating itself exactly. Quasi-periodic orbits are also closed loops but resemble necklaces, where the 'beads' are bunched together. This bunching represents the circumference of the loop expanding and contracting such that the particles do not repeat their trajectories exactly, yet still exhibit a regular pattern to their orbits.

What happens when $K > 0$? The fixed-point attractor $P = 0$, $\theta = \pi$ remains stable for $0 \leqslant K < 4$. For $K \geqslant 4$, this fixed-point attractor is unstable. In contrast, the fixed-point attractor $P = 0$, $\theta = 0$ is unstable whenever $K > 0$.

Consider $K = 0.5$, corresponding to light external forcing (Figure 3.11). Periodic orbits look like ellipses. These structures are typically called **islands**, as in islands of stability. For instance, a number of stable orbits exist for $-1 \leqslant \theta \leqslant 1$. Meanwhile, the quasi-periodic orbits now look like necklaces where the beads are wavy (e.g., $1.5 \leqslant \theta \leqslant 3.0$). These are all effects from the nonlinearity due to the the external

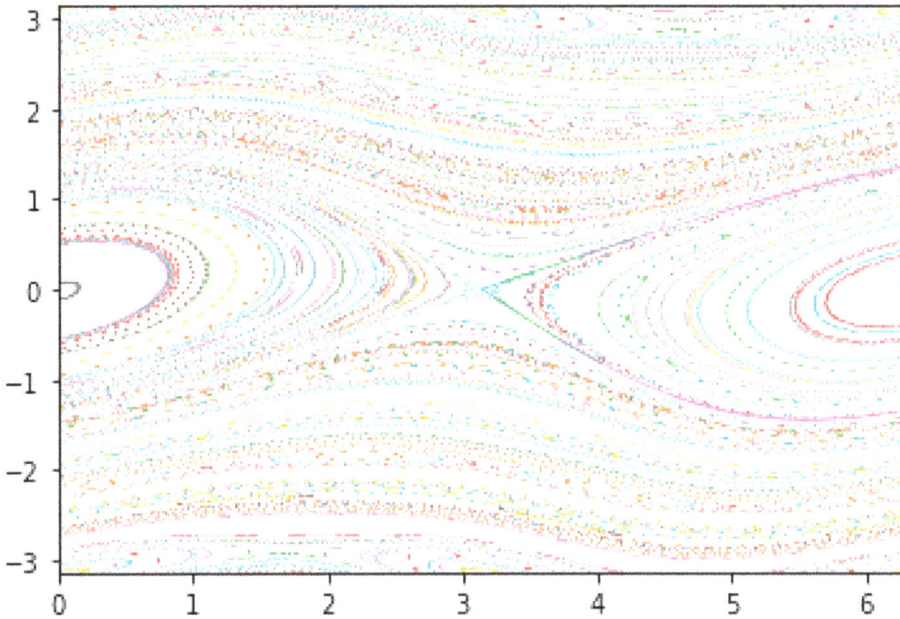

Figure 3.11. The Chirikov map plotting θ versus P for $K = 0.5$ displaying periodic and quasi-periodic orbits.

forcing term involving K. There is an unstable fixed-point attractor at $P = 0$, $\theta = \pi$, forming a saddle point—all trajectories move away from this point.

At $K = 0.971\ 635$, figure 3.12 reveals several islands of periodic orbits, very few quasi-periodic orbits, and several points appearing to have no pattern. The latter are aperiodic, the signature of chaotic behavior. A slight change in initial conditions, given a value of K above a critical value, leads quasi-periodic trajectories to become aperiodic. While it is conjectured that $K = 0.971\ 635$ is the value for the critical point, and there are some arguments supporting this, it has not been proven [5]. Notice that the saddle point attractor at $P = 0$, $\theta = \pi$ now is a point surrounded by aperiodic orbits.

Figure 3.13 displays the behavior of the Chirikov map for $K = 5.0$, a case of strong external forcing yielding a sea of chaotic trajectories surrounding islands of stability. One can say the same thing for figure 3.12 as well though it is less pronounced. This is a general feature of conservative maps with smooth generating functions exhibiting chaotic behavior.

Observe that the fixed-point attractor at $P = 0$, $\theta = 0$ is no longer stable (recall it's stable for $K \leqslant 4$). At $K = 4$, the attractor bifurcates into a period 2 loop attractor. As K increases, the period 2 attractor will undergo another doubling bifurcating into a period 4 attractor and so forth.

This process of periodic loop bifurcation continues for the Chirikov standard map, eventually producing a period 3 attractor and so forth leading to a sequence of periodic loop attractors of periods 1, 2, 3, etc. Each of these sequences is marked by a structure of islands of stability surrounding islands of stability. Such self-similar structures occur for all periodic loop attractors for conservative dynamical systems

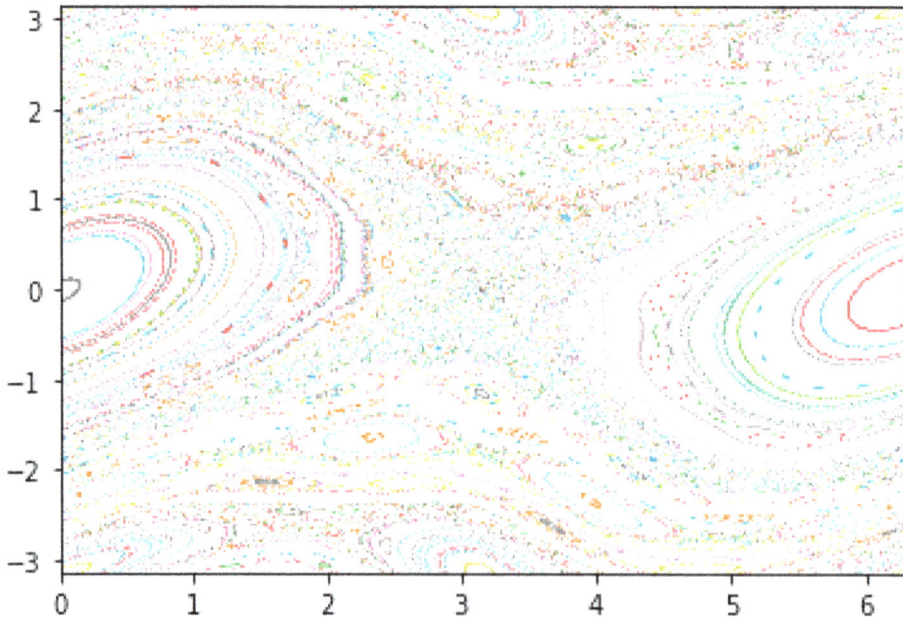

Figure 3.12. The Chirikov map plotting θ versus P for $K = 0.971\,635$, where some aperiodic orbits appear among the periodic and quasi-periodic orbits.

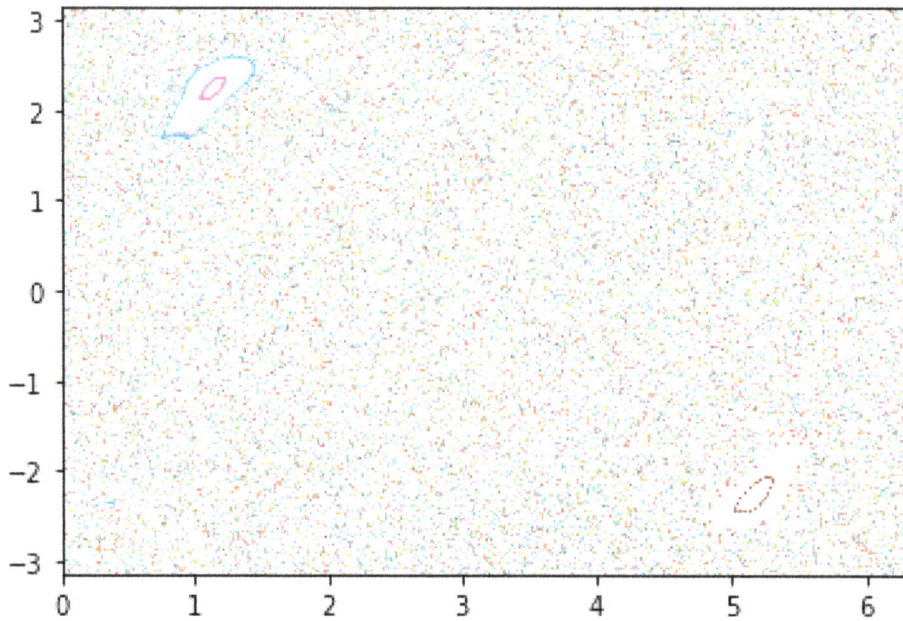

Figure 3.13. The Chirikov map plotting θ versus P for $K = 5.0$. There are always islands of stable periodic trajectories among the chaotic, aperiodic trajectories.

with smooth generating functions. This is the conservative dynamical system counterpart to strange attractors (section 4.1) in dissipative systems. These island-within-island structures repeat on smaller and smaller scales. Sometimes fractal structure (section 4.1) can appear in conservative as well as dissipative models; nevertheless, fractal structures in the former models aren't associated with strange attractors.

3.3 Observations

Having examined the Chirikov map in some detail, compare this behavior with figures 3.8 and 3.9. Just as for the Chirikov map, the the logistic map has many stable periodic orbits in the midst of a sea of aperiodic orbits. This is just as expected for area-preserving nonlinear maps with smooth generating functions. Furthermore, there are infinitely many periodic orbits. Both the logistic and Chirikov maps have periodic attractors of periods 1, 2, 3, etc. There is an infinity of periodic attractors because there is an infinitely of rational numbers corresponding to orbits of period n. Moreover, both the Chirikov and logistic maps have far (infinitely) more stable periodic than aperiodic orbits. Finally, the sequence of period doubling in the Chirikov standard map is characterized by the same Feigenbaum constant as with for the logistic map.

Stepping back, we can draw some observations from these examples. First, claims that chaos is ubiquitous are misleading unless suitably qualified. While nonlinearity is ubiquitous, chaotic dynamics isn't always present in nonlinear mathematical models. The logistic and Chirikov maps illustrate that chaotic dynamics only occurs when parameter magnitudes enable it.

Second, the randomness of the aperiodic orbits in the logistic and Chirikov maps, and the Lorenz system (equations (1.1)–(1.3)), is apparent randomness (section 1.1). Remember that the mathematical equations for all chaotic dynamical systems are deterministic. Deterministic dynamical systems are capable of exquisite and surprising order.

Third, as mentioned in the previous two chapters, these results have implications for predictability of chaotic systems: surprisingly, chaotic systems are far more predictable than the popularized discussions portray. For instance, nonlinear dynamical systems exhibit nonchaotic behavior for far more parameter settings than lead to chaotic behavior. Nonetheless, even when there is chaotic dynamics, a great deal of the intricately ordered behavior of these systems is predictable. The key issue with SDIC associated with chaotic dynamics is handling the growth in uncertainties (chapter 6).

References

[1] May R M 1976 Simple mathematical models with very complicated dynamics *Nature* **261** 459–67
[2] Feigenbaum M J 1978 Quantitative universality for a class of nonlinear transformations *J. Stat. Phys.* **19** 25–52

[3] Chirikov C V 1969 *Research concerning the theory of nonlinear resonance and stochasticity* CERN-Trans-71-40 Institute of Nuclear Physics, Novosibirsk http://cds.cern.ch/record/325497?ln=en

[4] Chirikov C V 1979 A universal instability of many-dimensional oscillator systems *Phys. Rep.* **52** 263–379

[5] Chirikov B and Shepelyansky D 2008 Chirikov standard map *Scholarpedia* **3** 3550

IOP Publishing

An Introduction to Chaotic Dynamics
Classical and quantum
Robert C Bishop

Chapter 4

Dissipative chaos and strange attractors

The previous chapter introduced the phenomenology of chaos using conservative or Hamiltonian (section 1.6) dynamical systems as models. This chapter explores strange attractors, chaos in dissipative systems, and compares these and conservative systems.

4.1 Strange attractors

A fifth attractor (section 1.6) of note is associated with dissipative systems, so-called **strange attractors**, mathematical objects having an infinite number of layers of repetitive structure. Informally, take any portion of a strange attractor, magnify it, and the same structure will appear. Repeat this process again and again and the identical structure will appear ad infinitum. This kind of structure is a **fractal**, an object having fractional dimension.

The usual conception of dimensionality can be generalized.[1] Start with a square. Fill this original square with smaller squares each having an edge length ε. Counting the number of small squares needed to completely fill the area inside the original square would yield $N(\varepsilon)$. Repeating this process with smaller and smaller squares of length ε yields a larger $N(\varepsilon)$. In the limit as $\varepsilon \to 0$, the ratio $\ln N(\varepsilon)/\ln(1/\varepsilon) = 2$ just as expected for a two-dimensional square. Similarly, the same exercise of filling a large three-dimensional cube with smaller cubes of edge length ε leads, in the limit $\varepsilon \to 0$, to the expected dimension three.

Applying this generalization of dimensionality to the geometric structure of strange attractors yields noninteger dimensions in the limit as $\varepsilon \to 0$. If the dimensionality of a curve is 1.74, for example, then it occupies more space than a line but less than a plane or surface. Similarly, if an object has a dimensionality of

---

[1] There is no agreement on how to define fractal dimensions. What follows is a simplified square or box counting method related to Hausdorff dimensionality [1].

doi:10.1088/978-0-7503-6453-9ch4

2.33, then it occupies more space than a surface but less than a three-dimensional volume.

This is the usual type of fractal discussed in popular literature, where a geometric pattern, say, is repeated exactly upon successive magnifications. The other kind of fractal is where only statistical characteristics are repeated on successive magnifications. Often, the more physically relevant fractals are those where statistical properties are repeated at different length scales.

Solutions to nonlinear mathematical equations, or the data from a time series generated by an experiment, exhibiting a strange attractor indicate that the behavior of the system under study is dissipative. Indeed, chaos in dissipative systems often is associated with strange attractors.

Nevertheless, caution is in order here. Detecting a fractal in the state space representation of a time series doesn't necessarily imply deterministic dynamics. Consider the unit interval [0, 1]. Subtract the middle third of this interval producing a new set. Next, remove the middle third of the remaining intervals to produce a new set of intervals. Continuing this procedure, in the limit leads to the Cantor set. In contrast to the original unit interval, having dimension 1, the Cantor set has dimension ~0.631. It is possible to produce a nondeterministic version of this procedure, therefore if a fractal produced in this fashion was detected, it would be unsafe to infer that a deterministic process produced the fractal.

This same example illustrates that detecting a fractal in a state space representation of a time series doesn't guarantee the system producing it is chaotic: dissipative chaotic systems will have fractal attractors, but not all dissipative systems having fractal attractors will be chaotic. For instance, the equation for a quasiperiodically-forced damped pendulum can exhibit a nonchaotic strange attractor [2].

4.2 The Lorenz system

Lorenz's 1963 system of equations (1.1)–(1.3) [3] is the earliest published example of a strange attractor even if this wasn't fully realized at the time. Figure 4.1 is perhaps the most identifiable example of chaotic dynamics.[2]

Strange attractors in state space allow trajectories to remain within a bounded region of the space by folding and intertwining with one another without intersecting or repeating themselves. Once a trajectory in state space wanders too close to the Lorenz strange attractor, it enters the basin of attraction and spends the rest of its future winding around the attractor.

Lorenz's system has a symmetry $(X, Y, Z) \to (-X, -Y, Z)$, and has three possible steady-state solutions, where no convection takes place. One is $X = Y = Z = 0$. The other two steady-state solutions are the centers of the two 'butterfly wings,' $X = Y = \pm\sqrt{b(r-1)}$, $Z = r - 1$ for $r > 1$ ($r = 1$ is the critical value for the onset of convection). Convection is always stable so long as $\sigma < b + 1$,

---

[2] It is important to note that figure 4.1 is a projection onto the X, Z-plane (Y held constant). Although the projection gives the appearance of trajectories intersecting themselves, in three dimensions trajectories never cross. Any orbits starting on the Z-axis remain on it.

Figure 4.1. The Lorenz 'butterfly' attractor for equations (1.1)–(1.3). Courtesy of Wikimedia Commons.

while for $\sigma > b + 1$ steady convection will be unstable for sufficiently high Rayleigh number. Lorenz was able to show that all orbits eventually enter a spherical region defined by $X^2 + Y^2 + Z^2 < R$ for some constant R, a clear indication that the dynamical system is dissipative.

Considering this system with parameter magnitudes $\sigma = 1.0$ and $b = 8/3$, the Jacobian matrix can be examined to determine stability of the flow:

$$
\begin{bmatrix}
-\sigma & \sigma & 0 \\
r & -1 & 0 \\
0 & 0 & -b
\end{bmatrix}
\tag{4.1}
$$

The eigenvalues are all negative for $0 < r < 1$, and $X = Y = Z = 0$ is the only attractor of the system (a fixed point attractor). $X = Y = Z = 0$ turns out to be globally stable, where every orbit approaches the stable fixed point as $t \to \infty$. Once r passes the critical value, however, the fixed point attractor becomes nonstable, meaning that $(0, 0, 0)$ is neither stable nor unstable, while $X = Y = \pm\sqrt{b(r - 1)}$, $Z = r - 1$ become stable fixed point attractors forming a pitchfork bifurcation from the point $X = Y = Z = 0$. As r increases, orbits in state space will spiral around one or the other of the latter fixed point attractors.

The magnitude of r in equation (1.2) measures the effect of the temperature differences between the upper and lower surfaces of the fluid layer: the larger the temperature difference between the warmer lower boundary and the cooler upper boundary, the stronger the buoyancy force due to warmer fluid rising in the presence of gravity. Dissipation outcompetes the convection of heat until the temperature difference is large enough so that $r = 1$. Beyond this critical value, convective flow is regular until $r \simeq 13.96$, where transient behavior resembling chaos sets in, though it doesn't attract orbits and dies out with the orbits eventually falling into the basin of attraction of one or the other stable fixed points ($X = Y = \pm\sqrt{b(r-1)}$). Indeed, for magnitudes of $13.96 \leqslant r < 24.74$, an infinity of stable periodic orbits are generated. Nonetheless, which attractor basin a trajectory eventually falls into is sensitively dependent to the initial conditions. As r is increased, trajectories spend increasing time exhibiting chaotic behavior before falling into one of the basins of attraction of the stable fixed points.

When $r \simeq 24.06$, a fractal chaotic attractor appears—a strange attractor—while at $r \simeq 24.74$ the two stable fixed points become nonstable and only the strange attractor remains for the dynamics [4]. The Lorenz strange attractor has a fractional dimension of 2.0627160 [5]. The Lorenz strange attractor contains a countable infinity of nonstable periodic orbits and an uncountable infinity of aperiodic orbits. There also is an uncountable infinity of orbits that will eventually terminate at the origin despite the fact that it is a nonstable point [6].

Further increasing r leads to more complex chaotic dynamic behavior, yet large enough r leads to a return to periodic behavior. For instance, stable periodic orbits exist in the range $99.524 < r < 100.795$. All computed trajectories are attracted to one of the stable orbits. A period-doubling bifurcation takes place between as $r \approx 99.98$. Interestingly, as r decreases towards 99.524, a sequence of period-doubling bifurcations occur with good evidence this is an infinite sequence of doublings [6]. Chaotic behavior appears again for $r = 100.8$. A similar period-doubling sequence occurs in the parameter range $145 < r < 166$ with chaotic behavior re-emerging at $r = 166.3$. When $r > 313$, the Lorenz system has a global stable limit cycle attractor.[3] Figure 4.1 illustrates the Lorenz attractor for $r = 28$. The slightest change in initial conditions for $r \geqslant 24.74$ leads to exponential growth in the uncertainty for any small neighborhood around many initial conditions. This translates into trajectories issuing from neighboring points around the attractor spreading exponentially apart from each other. Furthermore, slight differences in initial conditions determine the exact sequence of loops around one wing or the other of the attractor a particular trajectory makes as well as how long it spends orbiting one wing before switching to the other.

The Lorenz attractor illustrates the stretching and folding dynamics in a visually striking way. The stretching of trajectories easily can be imagined as orbits start out at neighboring points but soon end up on different wings of the attractor. The folding of the orbits, which can be seen as the wings converge back towards the center, builds up the multiple repeating layers of the strange attractor. This stretching and folding dynamic is characteristic of dissipative nonlinear dynamical systems exhibiting chaos, such as the Rössler attractor [7, 8] (see cover). It is a general feature of locally unstable yet globally bounded flows that trajectories issuing from within any open ball of initial conditions will be stretched and folded.

---

[3] Using a Poincaré section of surface, the return map for the Lorenz system forms a unimodal map with properties similar to the logistic map [6].

For strange attractors, such as Lorenz's, there are periodic orbits (both stable and nonstable) along with the aperiodic orbits. In fact, there is a countable infinity of periodic orbits, but these orbits are nonstable: Allowing time to run long enough, each aperiodic (chaotic) trajectory will return to the neighborhood of its initial position though with no discernible period. Perturbing the initial condition slightly can cause the perturbed orbit to close on itself forming a periodic orbit. Disturb the initial condition for the latter periodic orbit and it will form an aperiodic orbit.[4]

4.3 Prefractals versus fractals[5]

Fractals, such as strange attractors, have repeating structure on every scale in state space. In contrast, a **prefractal** only has two or three scales of repetitive self-similar structure before the self-similarity—even statistical self-similarity—breaks down, though they still have fractal dimension. While fractals arise in mathematical models, actual-world systems only exhibit self-similar structure on two or three scales (e.g., [11]).

Do strange attractors exist in actual-world systems? Perhaps strange attractors are artifacts of our mathematical models. This relates directly to the faithful model framework (section 1.9): if there are only prefractals in the actual world, but our dissipative dynamical systems harbor strange attractors, what's the relationship between our models and actual-world systems? How much mathematical structure captures physical structure? The fractal features of many of our models appear to be false of the target systems, yet these models are useful for helping scientists locate interesting dynamics of target systems. Are our models only useful fictions?

One reason to be cautious about this conclusion is that the prefractal character of the analyzed data sets [11] could be an artifact of the way data is digitized before analysis (e.g., analog-to-digital conversion must take place before data analysis can begin). Reducing real number valued data to a finite string possibly destroys fractal structure. Perhaps infinitely self-similar structures of strange attractors in our models might not be a bad approximation.

However, likely there is more than a data digitizing problem. The idea that physical systems have self-repeating structures 'all the way down' potentially is blocked by the transition from the macroscopic scale to the microscopic, where quantum mechanics (QM) becomes relevant. In the quantum realm, things change so dramatically state spaces cannot support strange attractors (chapter 8; see also [12]). This transition implies limits to our inferences based on models using non-quantum state spaces.

Employing obviously false chaos models might be justified pragmatically because the infinitely intricate structure of strange attractors is the result of stretching and folding mechanisms, where the points of interest in state space are invariant under these mechanisms. If so, then strange attractors exhibiting infinite self-similar structure in our models would be an indicator of some stretching and folding

---

[4] This interleaving of aperiodic and nonstable periodic orbits was implicit in Birkhoff's [9] but remained largely unexplored.

[5] For a discussion of philosophical implications of prefractals versus fractals, see [10], section 5.

process. The infinite structure might be extra geometric baggage, but robust properties of our models, such as period-doubling sequences, onset of chaos, and so forth, track with actual-world system behaviors. The stretching and folding processes giving rise to chaotic dynamics are important for characterizing systems as chaotic (chapter 5).

Although it's common to appeal to strange attractors as explaining chaos in actual-world systems in the literature, this may represent misplaced emphasis on what's doing the genuine explanatory work. For the sake of argument, suppose one appealed to strange attractors in our models or in state space reconstruction techniques as explaining chaotic behavior. Is this evidence that a strange attractor exists in the space of physical possibilities of target system behaviors? Given issues with the faithful model framework, even if detecting the presence of a strange attractor in the model state space was both a necessary and sufficient condition for the model being chaotic, this wouldn't amount to an explanation of chaotic behavior in the target system. First, strange attractors are mathematical objects in state space. This isn't the same as an actual system behaving as if there is a strange attractor in the physical space of its activity. Although trajectories in a state space are useful for gaining information about target systems (via the faithful model framework), this is different from trajectories developed by looking at how an actual system's properties change with respect to time. Even a system trajectory in state space spiraling ever closer to a strange attractor doesn't imply the target system's behavior in physical space is approaching closer to a strange attractor in physical space.[6]

Second, the presence of a strange attractor still is only an indicator of chaos, not an explanation for why a target system is exhibiting behavior. We still need to appeal to physical processes and interactions causing the chaotic dynamics such as stretching and folding dynamics (section 1.9 and 2.2).

4.4 Comparing dissipative and conservative chaotic systems

Although there are more dissipative than conservative systems in nature, as noted in section 3.2, fractal structure can be found in the Chirikov map exhibiting chaotic dynamics. Indeed, both the logistic and Chirikov maps have a bifurcation pattern leading to chaos (e.g., period-doubling sequences followed by period trebling sequences and so forth) just as found in the Lorenz and other dissipative systems. The result of these sequences of bifurcation is statistically self-similar fractal structure in the logistic and Chirikov maps on smaller and smaller scales (e.g., Section 3.3).

Similarly to the Lorenz and other dissipative systems, conservative dynamical system state space points leading to chaotic orbits, as in the logistic and Chirikov maps, are surrounded by points leading to stable and nonstable periodic orbits. There is an infinity of periodic orbits with more stable and nonstable periodic than there are aperiodic orbits in both dissipative and conservative systems.

---

[6] Except possibly under the perfect model framework.

As described in chapter 1, the logistic, Chirikov, and Lorenz dynamical systems are all deterministic and their chaotic dynamics only exhibits apparent randomness. But also notice the important role of parameters illustrated by all forms of chaos. While it's the parameter magnitudes that condition when chaotic dynamics appears, there are always vastly more magnitudes leading to stable than to chaotic behavior for all nonlinear dynamical systems.

The damped driven pendulum illustrates the important role parameters play in chaotic dynamics observed in both mathematical models and physical systems:

$$\frac{d^2\theta}{dt^2} + \frac{b}{m}\frac{d\theta}{dt} + \frac{g}{r}\sin\theta = \frac{F_0\cos(\omega t)}{mr}, \tag{4.2}$$

where θ is the angular position, b is the working coefficient of friction at the pivot point, m is the mass of the pendulum (assumed concentrated in the bob at the end), g is the acceleration due to gravity, r is the pendulum length, F_0 is the amplitude of the driving force, and ω is the angular driving frequency.[7] Examining equation (4.2), it's easy to show that this system exhibits chaos for some combinations of the magnitudes of friction, and amplitude and frequency of driving force. The parameter magnitudes, which determine the strength of the contributions of the nonlinear effects, are crucial determinants for the presence of chaotic dynamics in any deterministic nonlinear system, mathematical or actual.

References

[1] Hausdorff F 1918 Dimension und äußeres maß *Math. Ann.* **79** 157–79
[2] Romeiras F J and Ott E 1987 Strange nonchaotic attractors of the damped pendulum with quasiperiodic forcing *Phys. Rev.* A **35** 4404
[3] Lorenz E N 1963 Deterministic nonperiodic flow *J. Atmos. Sci.* **20** 130–41
[4] Kaplan J L and Yorke J A 1979 Preturbulence: a regime observed in a fluid flow model of Lorenz *Commun. Math. Phys.* **67** 93–108
[5] Viswanath D 2004 The fractal property of the Lorenz attractor *Physica* D **190** 115–28
[6] Sparrow C 1982 *The Lorenz equations: Bifurcations, Chaos, and Strange Attractors* (New York: Springer) Applied Mathematical Sciences https://doi.org/10.1007/978-1-4612-5767-7
[7] Rössler O 1976 An equation for continuous chaos *Phys. Lett.* A **57** 397–8
[8] Letellier C and Rössler O 2016 Rössler attractor *Scholarpedia* **1** 1721
[9] Birkhof G 1927 On the periodic motions of dynamical systems *Acta Math.* **50** 359–79
[10] Bishop R C C 2024 *The Stanford Encyclopedia of Philosophy* (Winter 2024 Edition), E N Zalta and U Nodelman (eds.), https://plato.stanford.edu/archives/win2024/entries/chaos/
[11] Avnir D *et al* 1998 Is the geometry of nature fractal? *Science* **279** 39–40
[12] Bishop R C 2024 *The Physics of Emergence* 2nd ed (Bristol: IOP Publishing) https://doi.org/10.1088/978-0-7503-6367-9

---

[7] Note that if the driving force, $F(t)$, was a constant, it would drive the pendulum to an equilibrium position in the case where the amplitude is too small to overcome friction and gravity, or would drive the pendulum to periodic motion in the case where the amplitude is large enough to overcome friction and gravity.

Chapter 5

Challenges for defining chaos

Having explored the phenomenology of chaotic dynamics in the previous two chapters, we turn to the question of how to define chaos. After all, a definition is needed to identify a system as chaotic. This turns out to be much subtler than is usually indicated in the literature. Nevertheless some useful marks and definitions for chaotic behavior exist.

5.1 Qualitative considerations[1]

Given how much popularized discussions of chaos focus on unpredictability, it would be easy to mistake this as a defining mark of chaotic systems. Yet, unpredictability is neither necessary nor sufficient to distinguish chaos from non-chaotic behavior (see appendix A). Much attention has been given to aperiodicity and instability in the literature, as illustrated in the previous chapters. Nonetheless, the presence of aperiodic orbits by themselves is neither necessary nor sufficient to identify chaotic from other forms of dynamics. Exponential growth in uncertainties is often identified as a defining feature of chaos, but there are many models that display forms of exponential growth that clearly aren't chaotic.

Another feature of chaotic systems is the presence of stretching and folding mechanisms in the dynamics. Such mechanisms are related to nonlinearities and cause some trajectories to converge rapidly while causing others to diverge rapidly. Such stretching and folding mechanisms lead to the growth of uncertainties as trajectories issuing from nearby points in some small neighborhood of state space and the mixing and dramatic separation of trajectories as illustrated by the Lorenz attractor (Figure 4.1): some nearby orbits end up on one wing while others end up on the other wing rapidly. For the appropriate parameter magnitudes, equations (1.1)–(1.3) stretch and fold orbits generating chaotic dynamics. Or consider the logistic map (section 3.1). It stretches the interval and folds it from a straight line into a

---
[1] This section draws heavily from [1].

doi:10.1088/978-0-7503-6453-9ch5

parabola, where successive iterations repeat this stretching and folding process eventually producing chaotic dynamics for appropriate magnitudes of the parameter α.

Explosive growth in uncertainties is associated with the stretching of trajectories—the spreading out of trajectories from each other—while the folding confines all trajectories to a region of state space so no orbits to go off to infinity. The presence of appropriate stretching and folding mechanisms in the dynamics may be a necessary condition for chaos. However, the identification of such mechanisms in target systems can be rather tricky. On the one hand, for fluid systems several nonlinear mechanisms have been well explored as sources for stretching and folding. On the other hand, when dealing with a time series (e.g., the hourly price of Chicago Board of Trade hog futures), identifying possible nonlinear mechanisms involved in stretching and folding is difficult.

These subtleties aside, stretching and folding mechanisms lead to dynamics with attractors, hence focusing on such mechanisms for characterizing chaos appears fruitful. From a qualitative standpoint we have determinism, nonlinearity, stretching and folding dynamics, aperiodicity, and sensitive dependence on initial conditions (SDIC) as potentially relevant necessary properties for defining behavior fitting our intuitions for chaos.

5.2 Quantitative definitions of chaos[2]

A quantitative approach to defining chaos begins with distinguishing weak versus strong forms of SDIC. Growth in uncertainties often is characterized using the **propagator** $J(\vec{x}, \Delta t)$, where trajectories $\vec{x}(t + \Delta t) = J(\vec{x}, \Delta t)$. To illustrate for two dimensions, consider an infinitesimal circle in state space representing the uncertainty in $\vec{x}(0)$ of radius $\delta(\vec{r})$. The action of $J(\vec{x}, \Delta t)$ transforms the circle to an ellipse of semimajor axis $\vec{r}_1 = \delta \vec{r}(0)e^{\varepsilon_1 t}$ and semiminor axis $\delta \vec{r}(0)e^{\varepsilon_2 t}$ after some amount of time as illustrated in figure 5.1.[3]

Let $\vec{x}(0)$ and $\vec{y}(0)$ be the initial conditions for two neighboring trajectories. **Weak sensitive dependence** (WSD) is defined as

Definition 5.1. [WSD]
$\exists \varepsilon > 0$ such that $\forall \vec{x}(0)$ and $\forall \delta > 0$ $\exists t > 0$ and $\exists \vec{y}(0)$ such that $|\vec{x}(0) - \vec{y}(0)| < \delta$ and $|\vec{x}(t) - \vec{y}(t)| > \varepsilon$.

Any trajectory issuing from $\vec{y}(0)$ will eventually diverge by ε from the trajectory issuing from $\vec{x}(0)$ no matter how small δ is. However, note that WSD doesn't specify the rate of this divergence—it's compatible with linear growth in uncertainty. Nor does it specify the number of points in the neighborhood of $\vec{x}(0)$ giving rise to diverging trajectories.[4]

---

[2] This section draws heavily from [1].
[3] In the limit of large time (i.e., infinitely long times), the factors ε_i may be replaced global Lyapunov exponents (Appendix C).
[4] It could be a set of measure zero.

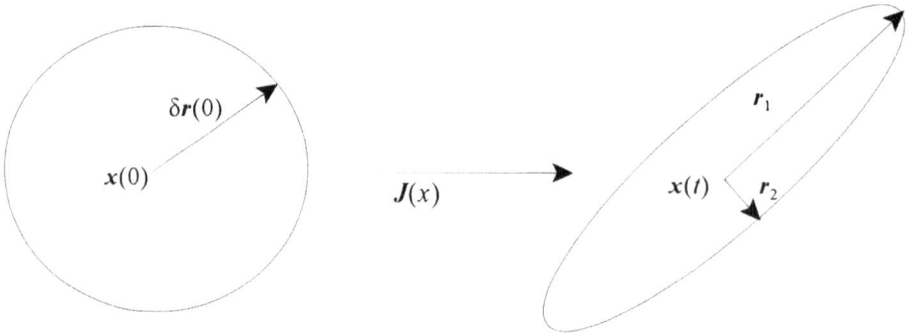

Figure 5.1. Evolution of an infinitesimal neighborhood of radius $\delta \vec{r}$ under the action of the propagator $J(\vec{x}, \Delta t)$. After some time t, the circle evolves into an ellipse of semimajor axis $\vec{r}_1 = \delta \vec{r}(0)e^{\lambda_1 t}$ and semiminor axis $\delta \vec{r}(0)e^{\lambda_2 t}$.

In the literature, SDIC usually is characterized by **strong sensitive dependence** (SD):

Definition 5.2. [SD]

$\exists \lambda$ such that for almost all points $\vec{x}(0)$ except a set of measure zero, $\forall \delta > 0$ $\exists t > 0$ such that for all points $\vec{y}(0)$ in a neighborhood δ around $\vec{x}(0)$ except for a set of measure zero, $|\vec{x}(0) - \vec{y}(0)| < \delta$ and $|\vec{x}(t) - \vec{y}(t)| \approx |\vec{x}(0) - \vec{y}(0)|e^{\lambda t}$.

Usually, λ is interpreted as the largest global Lyapunov exponent (Appendix C) and is assumed to represent the average rate of growth of uncertainty for trajectories originating from some small neighborhood centered around $\vec{x}(0)$. Exponential growth (convergence) is implied when $\lambda > 0$ ($\lambda < 0$). Such exponential growth will saturate at some time t for systems bounded in space and momentum.

Several definitions for chaos are formulated with discrete models and then generalized to continuous models. Either an iterated map, such as the logistic map, or a Poincaré section of surface can provide discrete models for this definitional strategy. Let f be a smooth map defined on the state space S. Furthermore, let K be a set of points and suppose that $f: K \to f(K)$.[5] Now Robert Devaney's [2] influential chaos definition can be stated:

Definition 5.3. [Chaos$_D$]

1. f satisfies WSD on K,
2. The set of periodic orbits of f are dense in K,
3. f is topologically transitive on K.

---

[5] The set K is an invariant of f if $f(K) = K$.

Topological transitivity is the following notion: let S be a space endowed with a metric and $f: S \to S$. The dynamical system f is topologically transitive if for every pair of nonempty open sets U and V in S, there is a nonnegative number N such that $f^N(U) \cap V \neq \varnothing$. Roughly, this means that some trajectory initiating from U eventually visits V. Topological transitivity guarantees trajectories starting from points in U will eventually explore all of S.

Mathematicians typically favor Chaos$_D$. The definition can look counterintuitive focusing on periodic orbits rather than aperiodicity, when, as we've seen, the latter are relevant for chaotic dynamics. However, if the set of nonstable periodic points is dense in K, then given (3) the abundance of aperiodic orbits characteristic of chaos is guaranteed. This is behavior we expect for dynamical systems exhibiting chaos. In 1992 Banks *et al* [3] demonstrated that conditions (2) and (3) imply (1) if K has an infinite number of elements.[6] Some have argued that (2) is not necessary for characterizing chaos (e.g., [4]).

Motivated by the Smale horseshoe [5], another possible chaos definition is

Definition 5.4. [Chaos$_h$]
A discrete function f is chaotic if, for some iteration $n \geqslant 1$, it transforms the unit interval I into a horseshoe.

For the Smale horseshoe, beginning with the unit square, stretch it by more than a factor of 2 in the y direction. Next, compress it by more than a factor of two in the x direction. Follow these operations by folding the resulting rectangle over and laying it back onto the square. The construction overlaps and leaves the middle and vertical edges of the initial unit square uncovered. Repeating these stretching and folding operations n times produces the Smale attractor.

One virtue of this definition is that Chaos$_h$ implies Chaos$_D$. Another is that Chaos$_h$ leads to SD whereas Chaos$_D$ doesn't. A third virtue is that this definition involves a stretching and folding mechanism. Nonetheless, a significant disadvantage is that Chaos$_h$ cannot be applied to invertible functions, the kinds of functions characteristic of many systems exhibiting Hamiltonian chaos. Ideally, a single definition should apply to both conservative and dissipative chaotic dynamics.

Another definition related to Chaos$_h$ is

Definition 5.5. [Chaos$_{te}$]
A discrete function f is chaotic just in case it exhibits topological entropy.

For the definition of topological entropy, see appendix B. Given a neighborhood N around $\vec{x}(0)$ less than a positive number ε away from each other, after n iterations of f the trajectories issuing forth from N will differ by at least ε. Moreover, an

---

[6] This result doesn't hold for sets with a finite number of elements.

increasing number of trajectories will differ by at least ε as n increases. For maps, Chaos$_h$ implies Chaos$_{te}$. But the latter is often easier to use to prove theorems.

The most popular candidate definition for chaos among physicists and other scientists is

Definition 5.6. [Chaos$_\lambda$]

A discrete function f is chaotic if it has a positive global Lyapunov exponent (Appendix C).

SD is the base for this definition, where positivity is defined as $\lambda > 0$ for almost all points in the specified set S. Computing global Lyapunov exponents for mathematical models often is advantageous, though it can be difficult to calculate these exponents for time series data generated by physical systems.

Global Lyapunov exponents (Appendix C) are usually thought of as parameters characterizing average growth rates of any uncertainty. These average rates are estimated over many initial conditions for each dimension of the system. As such, these exponents indicate the stretching or divergence rate per time step of a function averaged over a set of trajectories—the rapid divergence of neighboring trajectories away from each other due to the nonlinearities in the mathematical model.

To illustrate the usefulness of global Lyapunov exponents as potential diagnostics, compare the values of the Lyapunov exponents for the logistic map with its bifurcation diagram in figure 5.2. Observe that the sign of the global Lyapunov exponent tracks the bifurcation behavior precisely indicating when the logistic map is exhibiting chaotic behavior.[7] Also notice that there is a larger set of parameter magnitudes leading to stable behavior compared to chaotic.

One advantage of global Lyapunov exponents is that they provide a convenient measure of the growth in the uncertainty giving some insight into system predictability. A drawback is that these exponents only estimate growth for infinitesimal uncertainties (Appendix C). A more applicable measure of system unpredictability is given by the doubling rate 2^t, a measure of when the uncertainty grows to twice that in the initial data as t increases. In linear systems, uncertainty growth takes a very long time to reach twice that of the initial data. In nonlinear systems, exhibiting chaotic behavior, the doubling rate is $2^{\lambda t}$, where growth in uncertainty is thought to be a pure exponential with exponent λ.

Another definition somewhat popular with physicists and some mathematicians is that of Li and Yorke [6]:

Definition 5.7. [Chaos$_{LY}$]

Let J be an interval and let $f : J \to J$ be continuous. Assume there is a point $a \in J$ for which the points $b = f(a)$, $c = f^2(a)$, and $d = f^3(a)$, where $d \leq a < b < c$ (or $d \geq a > b > c$). Then,

---

[7] Such tracking is always the case for chaotic mathematical models.

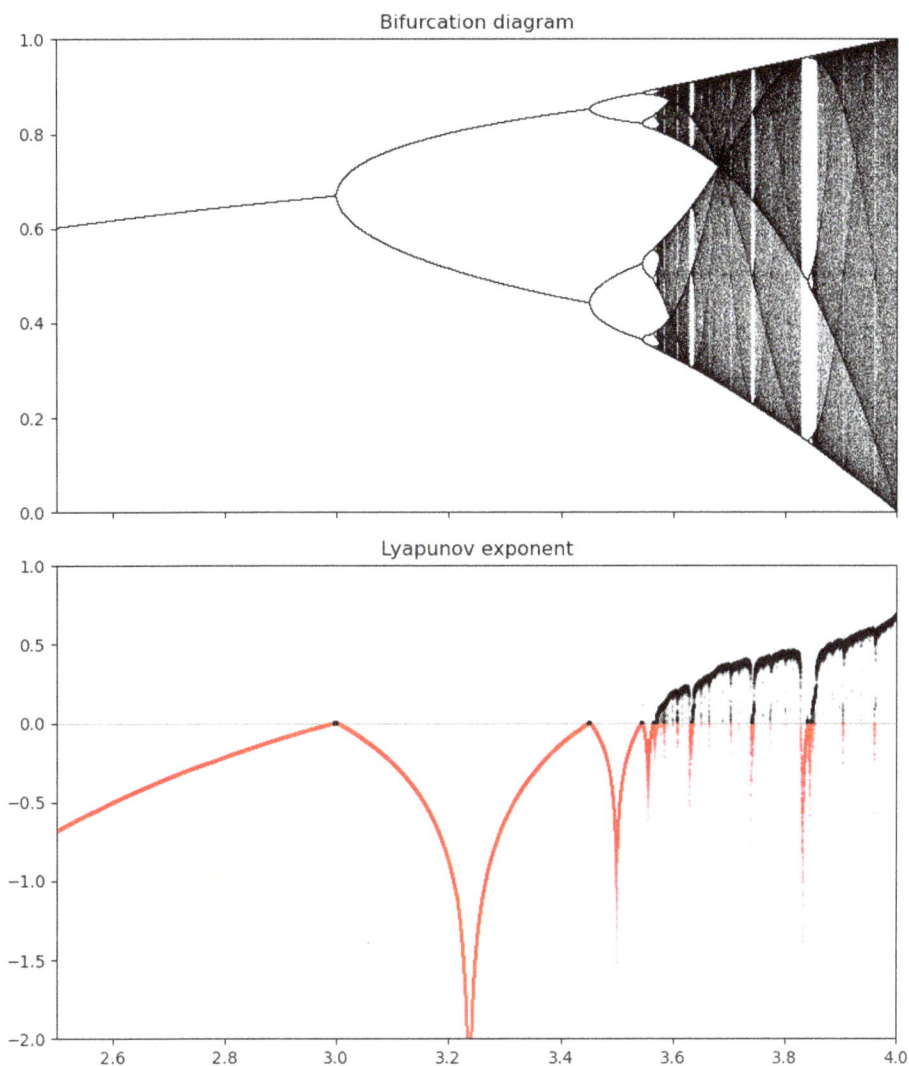

Figure 5.2. A plot of the values of the global Lyapunov exponent for the logistic map for α ranging from 2.5 to 4.0 at the bottom with the corresponding bifurcation diagram at the top. For the Lyapunov exponent values, black is negative while red is positive.

1. For every $k = 1, 2, 3, \ldots$ there is a periodic point in J having period k.
2. There is an uncountable set $P \subset J$ (containing no periodic points), satisfying the following conditions:
    (a) For every $x, y \in P$ with $x \neq y$, $\limsup_{n \to 0} |f^n(x) - f^n(y)| > 0$ and $\liminf_{n \to 0} |f^n(x) - f^n(y)| = 0$.
    (b) For every $x \in P$ and periodic point $y \in J$, $\limsup_{n \to 0} |f^n(x) - f^n(y)| > 0$.

Whenever there is a periodic point of period three, these conditions will be satisfied. However, Chaos_{LY} only implies WSD. Moreover, Chaos_D implies Chaos_{LY} but not vice versa.

A somewhat popular approach to defining chaos is through **algorithmic complexity** (e.g., [7, 8]), the measure of the number of elementary operations required for the execution of an algorithm relative to the size of the posed problem. Roughly, if the time evolution of a system requires at least N bits of input information about the initial state to obtain N bits of output information about its future state, then it's algorithmically complex. The string of information describing the future state given the initial state cannot be compressed. This is equivalent to a dynamical system algorithm's incompressibility, the inability to shorten the number of operations needed to compute the result. Algorithmic complexity is motivated by information theory considerations since rapid growth in uncertainty is related to information loss.

It turns out that classical chaotic trajectories are always algorithmically complex. Nonetheless, by Brudno's theorem algorithmic complexity in the classical context is not distinct from having a positive global Lyapunov exponent, thus is equivalent to the Kolmogorov–Sinai entropy (Appendix A) for almost all trajectories [9]. Hence, algorithmic complexity cannot serve as an independent definition of chaos.

Finally, some authors argue that chaos can be characterized in terms of the ergodic hierarchy [10]. Consult appendix A for details on the ergodic hierarchy and the chaotic hierarchy that serves as a concrete proposal.

5.3 Counterexamples

All definitions in the preceding section are based on the study of dynamical systems, and attempt to capture the ideas aperiodicity, of rapid uncertainty growth, and predictability limits. Yet, there is no consensus in the literature on a definition for chaos. One reason is that different definitions are more useful or practical in some contexts than others (e.g., for proving theorems). A second reason is that there are numerous counterexamples to all the proposed definitions.

Consider a two-dimensional dynamical system with a disc of radius ε about the origin removed from the plane. Define a twist

$$\frac{dx}{dt} = - ry,$$
$$\frac{dt}{dt} = rx, \tag{5.1}$$

where $r = \sqrt{x^2 + y^2}$ and $x^2 + y^2 > \varepsilon^2$. Orbits issuing from two neighboring initial conditions will rotate about the origin at different angular velocities. This means equations (5.1) exhibits WSD. Moreover, every orbit is on an invariant circle with the center as the origin. These orbits form an almost periodic sequence ensuring that the time series generated for each solution of the dynamical system is almost periodic. While this dynamical system exhibits WSD, it has neither positive global

Lyapunov exponents nor any exponential growth of uncertainties. If the aim of a definition is to capture the essential features of a phenomenon, this system illustrates that WSD is too broad to distinguish chaos from other behaviors.

Problems also arise for SD. Consider a discrete dynamical system with $S = [0, \infty)$, the absolute value as a metric (i.e., as the function defining distance between two points) on $\mathbf{R}$, and a mapping $f: [(0, \infty) \rightarrow [0, \infty), f(x) = cx$, where the parameter $c > 1$. This is a dynamical system where all neighboring trajectories diverge exponentially fast, fulfilling SD. Nevertheless, the forward-time autocorrelation function of $f(x)$—a measure of the correlation of future states with past states —equals one. Past states are perfectly correlated with future states (as expected for a linear relationship); hence, there is no randomness even though there is exponential growth of uncertainty. Or consider the simple linear differential equation $dx/dt = x$, whose solution is $x = x_0 e^t$. Once again, SD is fulfilled, the global Lyapunov exponent is positive, and there is exponential loss of uncertainty. Yet, this dynamical system isn't chaotic, calling into question chaos_λ as a stand alone definition for chaotic dynamics as well.[8]

Furthermore, for these latter two cases, all trajectories accelerate off to infinity. In contrast, chaotic dynamics is usually characterized as being confined to some attractor, a strange attractor in the case of dissipative systems, or an energy surface perhaps in the case of Hamiltonian systems. Confinement need not be due to physical walls of a container. If, in the case of Hamiltonian chaos, the dynamics is confined to an energy surface (e.g., by the action of gravity), this surface could be spatially unbounded. Thus, at the very least some additional conditions are needed (e.g., guaranteeing trajectories in state space are dense and confined), raising questions about whether additional conditions are needed for definitions such as SD, Chaos_{te} or Chaos_λ. On the other hand, a well-studied example of a chaos, the Hénon system[9], while measure preserving, is unbounded. Perhaps the usual restriction of chaos definitions to bounded systems needs qualification.

Nevertheless, even confining ourselves to bounded systems, the presence of a positive global Lyapunov exponent is insufficient to distinguish a dynamical system uniquely as chaotic. Consider the following dynamical system:

$$\frac{dx}{dt} = - rye^t,$$
$$\frac{dy}{dt} = rxe^t,$$

(5.2)

where $r = \sqrt{x^2 + y^2}$ and $x^2 + y^2 > \varepsilon^2$. The solution to equations (5.2) can be written in terms of a one-sided shift function by evaluating it at equally spaced times $t_n = n \ln 2$:

---

[8] Note that the presence of a positive global Lyapunov exponent doesn't lead to a horseshoe in this example, so Chaos_h would disagree with SD and Chaos_λ about the presence of chaotic dynamics.
[9] In the literature, the Hénon system usually is called a map rather than a flow because it's a discrete system.

$$x(t_n) = x_0 \cos(r_0(2^n - 1)) - y_0 \sin(r_0(2^n - 1)),$$
$$y(t_n) = y_0 \cos(r_0(2^n - 1)) + x_0 \sin(r_0(2^n - 1)),$$

(5.3)

where $r_0 = \sqrt{x_0^2 + y_0^2}$. Equations (5.3) have zero autocorrelation and a positive global Lyapunov exponent because of their relationship to a one-sided shift. Nonetheless, there is no sensitive dependence, not even WSD. The fixed point $(x_0, y_0) = (0, 0)$ is a solution of system (5.2) for all time and all other solutions stay nearby this fixed point for all time.

There are dynamical systems exhibiting exponential growth in uncertainty, and which have no autocorrelation function, but nevertheless don't exhibit SDIC. For example,

$$\frac{dx}{dt} = -y\frac{du(z)}{dt},$$
$$\frac{dy}{dt} = x\frac{du(z)}{dt},$$
$$\frac{dy}{dt} = 1,$$

(5.4)

where $u(z) = \cos(A(e^Z - 1))\log(z + 1)$. The presence of the exponential function in the argument of the cosine guarantees exponential growth in uncertainty in the constant A in the argument of a bounded function. The trajectories of system (5.4) wander randomly in state space implying no autocorrelation function can be computed. Yet, the system doesn't exhibit SDIC.

What about systems that exhibit SDIC and generate dense periodic orbits but lacks topological transitivity? Let a dynamical system consist of periodic orbits of different sizes centered round rational points, where no two periodic orbits intersect each other. Consider arbitrarily small deviations in initial conditions. The nature of rational numbers implies this system ends up in completely different orbits and hence diverging trajectories with no given point ending up in any open set challenging definitions, such as Chaos_D, SD, Chaos_{te} or Chaos_λ.

Or consider horocyclic flows defined on a compact surface of negative curvature. A horocycle is a curve on a surface where all normals to the curve (the perpendicular geodesics) through a point on the curve are limiting parallel and all converge asymptotically on a single ideal point. For instance, the set of vectors orthogonal to a Euclidean circle tangent to a boundary in the upper half-plane pointing inward is a stable horocycle. Any aperiodic horocyclic flows on a compact surface of negative curvature are ergodic, implying these flows are dense in the state space and topologically mixing [11]. Therefore, such flows exhibit topological transitivity. Note that periodic horocyclic flows are not dense but the points initiating periodic orbits are.[10] Barbara Schapira has shown that these orbits don't diverge [12], hence the global Lyapunov exponents computed for such trajectories are zero. Such orbits

---

[10] There are very few invariant ergodic measures for these aperiodic flows, meaning some of these flows can escape to infinity. Here, I restrict attention to those having invariant measure.

don't even exhibit WSD, raising questions about $Chaos_D$, since conditions (2) and (3) for Devaney's definition imply WSD for infinite systems. On the other hand, the periodic flows can have a positive global Lyapunov exponent, but aren't dense in the subspace, also challenging SD, $Chaos_{te}$, and $Chaos_\lambda$.

One motivation for the chaotic hierarchy is that in principle definitions such as $Chaos_D$, SD, $Chaos_{te}$ or $Chaos_\lambda$ can be mapped onto the ergodic hierarchy (Appendix A). Nonetheless, as we have observed, these candidate definitions run into problems from counterexamples. Some, however, have argued for the property of strong mixing (Appendix A) as a necessary condition for chaos. But this runs into subtleties.

Consider KAM-type systems, which are integrable systems with small perturbations and can have regions of both regular and irregular (stochastic) dynamics [13], p 166ff, [14], p 229ff. For an integrable Hamiltonian—a function describing the energy of a system[11]—imagine a torus and follow a trajectory as it orbits the energy surfaces. Consider its Poincaré section of surface under small perturbation. Even though KAM-type systems are not ergodic—hence, possess no properties on the ergodic hierarchy—the circles in the surface of section representing rational winding numbers get destroyed. This implies that the state space gets separated into two distinct regions, meaning the invariant curves on the energy surfaces get confined to different portions of the state space and cannot cross over any invariant tori.[12] There are many examples of concrete instantiations of such systems having regions of both regular and aperiodic orbits. The latter spread across their restricted portion of state space (e.g., the Hénon system and Chirikov standard map). Yet, KAM-type systems are not even ergodic, let along strong mixing. One at least has to restrict the claims for strong mixing as a necessary (or sufficient) condition for chaos to only those regions in state space where irregular behavior exists.

On the other hand, settling on $Chaos_D$, or equivalently, strong mixing as the definition of chaotic dynamics (e.g., [15]), commits one to classifying a large range of linear dynamical systems as chaotic. Indeed, there is a burgeoning research field called 'linear chaos' (e.g., [16, 17]). In linear dynamics, a dynamical system is hypercyclic if there is an $x \subset S$, an infinite-dimensional vector space, such that the orbits of the dynamical system are dense in S. By Birkhoff's transitivity theorem [18] such a dynamical system has the property of being topologically transitive. Hence, by $Chaos_D$, any infinite-dimensional linear dynamical system that is hypercyclic is chaotic even though the system never exhibits SD but only WSD. Nevertheless, it isn't clear why these systems should be classified as chaotic given that if one waits long enough almost every linear dynamical system will grow uncertainties large enough to cause some form of limitations on predictions of future states though these linear systems lack other properties of chaos such as exponential growth in uncertainties and zero autocorrelation functions.

---

[11] Give an Hamiltonian describing the total energy of a system, the dynamics of the system can be investigated in detail.

[12] For discussion, see [10], p 680–5.

Finally, there are many examples of algorithmically complex systems that aren't chaotic. For instance, long randomly generated bit strings, no matter how they were obtained, are algorithmically complex but need not have any relationship to chaos. Or consider the dynamical system

$$\frac{dx}{dt} = -2yt,$$

$$\frac{dy}{dt} = 2xt,$$

whose solution is a linear combinations of sine and cosine functions. The average autocorrelation function is zero everywhere except at $(x = y = 0)$, so the behavior of solutions is random, but there is no SDIC (or even WSD) and no exponential growth in uncertainty. The solutions are algorithmically complex while exhibiting no chaotic dynamics.

5.4 Observations[13]

Intuitions in the literature are that chaotic dynamics should be characterized by SDIC, exponential growth in uncertainty, a positive global Lyapunov exponent, a dense set of periodic orbits, and an autocorrelation function that decays to zero rapidly among other possible distinguishing marks.

There are dynamical systems exhibiting SDIC but which have no decaying autocorrelation functions. There also are systems with decaying autocorrelation functions but which don't exhibit exponential growth in uncertainty or even SDIC. Stepping back, the results of the previous section may be summarized as follows:

- Systems exhibiting WSD can fail to have positive global Lyapunov exponents or exhibit exponential growth in uncertainty.
- Attempts to align definitions such as $Chaos_D$, SD, $Chaos_{te}$, or $Chaos_\lambda$ with the ergodic hierarchy yield examples with exponential divergence of trajectories yet all past states are perfectly correlated with any future states yielding no randomness associated with chaos.
- Some cases with positive global Lyapunov exponents have trajectories accelerating off to infinity raising questions about the importance of requiring confinement.
- Even for bounded systems, the presence of a positive global Lyapunov exponent may not lead to exponential growth in uncertainties calling into question $Chaos_\lambda$.
- Some systems exhibit SDIC and generate dense periodic orbits, but lack topological transitivity raising questions about $Chaos_D$, SD, $Chaos_{te}$, and $Chaos_\lambda$.
- KAM-type systems raise question about strong mixing as either a necessary or sufficient condition for chaos.
- $Chaos_D$ or strong mixing classify much linear behavior as chaotic.

---

[13] This section draws heavily from [1], where the definition $Chaos_{Dexp}$ was first proposed.

Evidently it's difficult to craft a definition identifying chaotic behavior that is immune to counterexamples.

In an attempt to improve on this situation, one might suggest a modification to Devaney's definition:

Definition 5.8. [Chaos$_{Dexp}$]
   1. f satisfies SD on K,
   2. The set of periodic orbits of f are dense in K,
   3. f is topologically transitive on K.

This definition has the virtue of all three conditions being independent of each other while clearly distinguishing linear from nonlinear behavior reserving chaotic dynamics for the latter. Our intuitions about exponential growth in uncertainties as well as a dense set of aperiodic orbits are preserved.

Chaos$_{Dexp}$ also acknowledges that WSD is too weak to uniquely distinguish chaotic dynamics.[14] Specifying SD plus a set of dense periodic points and topological transitivity rules out cases where there is (1) a positive global Lyapunov exponent and no randomness, (2) a positive global Lyapunov exponent and no exponential growth in uncertainties, (3) a set of dense periodic points and topological transitivity but no positive global Lyapunov exponent, (4) KAM-type systems, and (5) linear dynamics. Although likely not immune to some counterexamples, this definition gets closer to characterizing the phenomenology of chaotic dynamics that all of the definitions in the previous section are aiming for.

In practice, it's the average Lyapunov exponents that are calculated to demonstrate when exponential growth in uncertainties occurs (Appendix C). This means that in actual practice what really gets plugged into a definition such as chaos $_{Dexp}$ is an average global Lyapunov exponent. What one should use are local finite-time exponents when seeking to capture detailed information about the attractor.

This is important because global Lyapunov exponents only measure infinitesimal growth in uncertainties, but infinitesimal effects cannot grow to finite size in finite time. So long as amplification remains infinitesimal, chaotic systems would never exhibit any form of SDIC. Finite-time, local Lyapunov exponents are the appropriate measures for the growth of uncertainties associated with chaos.

Finally, as previous chapters have illustrated, mathematical models only exhibit chaotic behavior for particular combinations of parameter magnitudes. Although often ignored in discussions about defining chaos, sensitive dependence on parameter settings shouldn't be neglected.

---

[14] Devaney still cites WSD in his most recent [19], see p 61.

References

[1] Bishop R C C 2024 *The Stanford Encyclopedia of Philosophy* (Winter 2024 Edition), E N Zalta and U Nodelman (eds.), https://plato.stanford.edu/archives/win2024/entries/chaos/

[2] Devaney R L 1989 *Chaos and Fractals: The Mathematics Behind the Computer Graphics* **vol 39** (Providence, RI: American Mathematical Society) 1–24 pp

[3] Banks J *et al* 1992 On Devaney's definition of chaos *Am. Math. Mon.* **99** 332–4

[4] Robinson C 1998 *Dynamical Systems: Stability, Symbol Dynamics and Chaos* 2nd ed (Boca Raton, FL: CRC Press) https://doi.org/10.1201/9781482227871

[5] Smale S 1967 Differentiable dynamical systems *Bull. Am. Math. Soc.* **73** 747–817

[6] Li T-Y and Yorke J A 1975 Period three implies chaos *Am. Math. Mon.* **82** 985–92

[7] Crisanti A *et al* 1994 Applying algorithmic complexity to define chaos in the motion of complex systems *Phys. Rev. E* **50** 1959–67

[8] Ford K 1986 Chaos: solving the unsolvable, predicting the unpredictable! *Chaotic Dynamics and Fractals* ed M F Barnsley and S G Demko (Orlando, FL: Academic) 1–52

[9] Brudno A A 1978 The complexity of the trajectory of a dynamical system *Russ. Math. Surv.* **33** 197–8

[10] Berkovitz J *et al* 2006 The ergodic hierarchy, randomness and chaos *Stud. Hist. Phil. Mod. Phys.* **37** 661–91

[11] Frustenberg H 1973 The unique ergodigity of the horocycle flow *Recent Advances in Topological Dynamics* ed A Beck (Berlin: Springer) 95–115

[12] Schapira B 2017 Dynamics of geodesic and horocyclic flows *Ergodic Theory and Negative Curvature* ed B Hasselblatt (Cham: Springer) 129–55

[13] Lichtenberg A J and Liebermann M A 1992 *Regular and Chaotic Dynamics* 2nd ed (New York: Springer) https://doi.org/10.1007/978-1-4757-2184-3

[14] Ott E 2002 *Chaos in Dynamical Systems* 2nd ed (Cambridge: Cambridge University Press)

[15] Werndl C 2009 What are the new implications of chaos for unpredictability? *Br. J. Phil. Sci.* **60** 195–220

[16] Gilmore C 2020 Linear dynamical systems *Ir. Math. Soc. Bull.* **86** 47–77

[17] Grosse-Erdmann K-G and Manguillot A P 2011 Linear chaos *Universitext* (London: Springer) https://doi.org/10.1007/978-1-4471-2170-1

[18] Birkhoff G 1922 Surface transformations and their dynamical applications *Acta Math.* **43** 1–119

[19] Devaney R 2021 *An Introduction to Chaotic Dynamical Systems* 3rd ed (New York: Chapman and Hall/CRC Press) https://doi.org/10.1201/9780429280801

IOP Publishing

An Introduction to Chaotic Dynamics
Classical and quantum
Robert C Bishop

Chapter 6

Implications for modeling and forecasting

Thus far, this book has focused primarily on chaotic behavior of mathematical models. The study of dynamical systems yields much insight into the phenomenon of chaotic behavior. But chaos also exists in the actual world. What is the relationship between chaos in our mathematical models and that observed in the actual world? How do scientists deal with the implications of chaotic dynamics for modelling and forecasting, where we face limits on prediction and must rely heavily on computers in modelling?

6.1 Chaos in the actual world

From driven damped pendulums to lasers [1], chaotic dynamics shows up in actual-world systems. Nevertheless, we've already encountered reasons to think that the relationship between our chaotic models and actual-world phenomena is subtle. For example, dissipative mathematical models can exhibit strange attractors with infinitely repeating structure, whereas systems in nature apparently only exhibit prefractal structure. We have also seen that global Lyapunov exponents present subtleties. While it may be relatively straightforward to calculate these exponents for simple mathematical systems, such as the Logistic map, it's more difficult to calculate them for dynamical systems such as Lorenz's equations (equations (1.1)–(1.3)). And it's much more difficult—perhaps even impossible sometimes—to measure global Lyapunov exponents for physical systems. As well, there are the issues presented by the faithful model framework (section 1.9).

On the other hand, recall the Feigenbaum constant computed for the magnitudes of the parameter α for successive bifurcations in the logistic map (section 3.1.2): $-4.699\,201\,6091...$. It is a delightful surprise that computing the Feigenbaum constant yields values very close to that of the Logistic map for period doubling in physical systems such as water, mercury, and liquid helium, or in chemical reactions such as the visually stunning Belousov–Zhabotinsky reaction, or in lasers,

doi:10.1088/978-0-7503-6453-9ch6

diodes, and transistors. However, we know there is no connection between the logistic map and these quite different physical systems.

Suppose we build a physical pendulum and set all the parameter magnitudes according to our model predictions for chaotic dynamics. Initially, the pendulum behaves chaotically as the model predicts (equation (4.2)). Over time, however, the friction at the pivot point heats up the materials leading to changes in the magnitude of the working coefficient of friction. Chaotic behavior can disappear because of the change in friction, whereas the working coefficient of friction in the mathematical model remains constant in time.[1] Likewise, the motor used to drive the pendulum heats up over time leading to variations in the amplitude and frequency of its driving action, whereas these parameters are treated as constant in time in the mathematical model. While chaotic behavior is persistent in our mathematical model, it's relatively short lived in actual physical pendulums due to variations in the physical magnitudes of key parameters.

The relationship of our models to physical systems requires care. One of the upshots of chapter 5 is that we don't have a universal definition for chaotic dynamics. Certainly, focusing on the stretching and folding dynamics of systems leading to sensitive dependence on initial conditions (SDIC) are important even if we have no simple surefire tests for these. While we may lack rigorous definitions for stretching and folding dynamics, scientists and engineers do have ways of investigating physical systems to understand the extent to which such dynamics are present. Stretching and folding dynamics as well as determinism are necessary conditions actual-world systems must possess if they are to be capable of behaving chaotically. Nevertheless, these conditions aren't sufficient to guarantee chaotic dynamics; we also must pay attention to the parameter magnitudes crucial to enabling chaos in systems.

When an actual-world systems such as the weather possess the necessary and sufficient conditions to exhibit chaotic behavior, what does that mean in practice? As the hurricane example in section 1.2 illustrates, any imprecision in the measurement of an earlier state of a weather system will lead to much larger uncertainty in the forecasted future weather state. This is an epistemic issue, a problem of our current versus projected future knowledge. Of course, the weather system will evolve based on the actual earlier state, not on our knowledge about that state. The physical behavior of the weather is dependent on the nature of the stretching and folding processes in its fluid dynamics, not the precision of our knowledge of the system and its earlier state. This mismatch between our knowledge and actual behavior is the apparent randomness we observe in actual-world systems (section 1.1).

Chaotic dynamics in weather or other actual-world systems implies limitations on our ability to forecast future system behaviors. Nonetheless, scientists extract useful forecasts from nonlinear systems such as the weather by paying attention to the lessons chaos is teaching us.

---

[1] Other subtleties about friction and surfaces are discussed in chapter 7.

6.2 Measurements, faithfulness, and the actual world

While advances also have been made in measurement accuracy in a wide variety of contexts, perfect measurement accuracy will never be achieved due to error reduction being a limiting process. Infinite precision is required to eliminate observational uncertainty for any chaotic model, and approaching such precision will exceed the information storage capacity of Universe [2].

Furthermore, any measurement apparatus introduces some small disturbance into systems being measured even at macroscopic scales, which cannot be ignored in nonlinear contexts though it's typical to assume such measurement disturbance can be ignored in classical physics settings (e.g., [3]). Hence, even if measurement inaccuracy could be reduced to zero, the act of measurement disturbs the True state of the system being observed. This uncertainty will be amplified (not to mention small disturbances due to minute fluctuations in gravity and electromagnetic fields owing to motions of massive charged objects such as vehicles). Any such measurement apparatus disturbances will be amplified by SDIC in any system exhibiting chaos. This returns us to the issues of the faithful model framework (section 1.9).

6.3 Numbers, computers, and modeling chaotic systems

Further modeling nuances arise due to our using numbers and computers. While integers are whole numbers, real numbers can have an infinitely long string of digits such as the value of π. Mathematical models always involve both kinds of numbers.

In contrast, physical measurements involve only integers. For instance, a person's weight might be measured on a scale as 77.107 kg, but this is an integer in disguise. Multiplying by 1000 yields the integer 77 107. Moreover, all numbers represented in digital computers and stored in memory devices are integers as well. The digital world is a world of integers only, while the world of mathematical models is a world of integer and real numbers.

Measurements are always bracketed by error bars, such as $77.107 \pm .005$ kg, quantifying the uncertainty. When scientists spell out the sources of uncertainty, they are specifying a model for the uncertainty (e.g., its sources, structure, etc). A very common model to use for uncertainty, as at least a first approximation, is a Gaussian or normal curve if there are no reasons to suspect that uncertainty is distributed asymmetrically about a mean value.

Here, we reach another contrast: the world of mathematics has no uncertainty. It is the realm of perfect models and exact precision. For example, when we set $\theta = \pi$ in the cosine function, π includes the infinite number of decimal places. Such mathematical models differ from the realm of digital computers, and perhaps from the actual world with its physical properties such as the temperature at O'Hare Airport.[2] I will call magnitudes of properties that exist independent of any measurement and uncertainty **ontic magnitudes** [4]. Over the decades, scientists

---

[2] One can ask some rather deep questions about the existence of True magnitudes of properties such as temperature, pressure, and wind speed at O'Hare Airport or any other spatial location.

have learned much about uncertainty and the degree to which ontic magnitudes exist in the actual world by exploring chaotic dynamics.

When it comes to studying systems on a computer, two types of uncertainty must be tracked. The first is the uncertainty in observations due to imperfect measurement apparatus among other sources as has been described previously.

The second type is uncertainty in our computer or model representations. If you input 77.0 kg into the computer to represent a person's weight but the ontic value is 77.107 kg, that is a representational error in the computer. Generally, representational uncertainty is more complex. Suppose we want to accurately estimate the average weight of rugby players in the world. It is not possible to measure the weight of every player, therefore we sample the weights of several players from different countries and average those to get an estimate. The more players we measure, the better the average estimate, but this estimate will always be an imperfect representation as reflected in error bars. Hence, our models for uncertainty will always contain some inaccuracy.

Consider how this plays out in weather forecasting. Scientists usually measure quantities such as temperature, pressure, and humidity in some region and these data are input into their forecasting models. Of course, there always is some uncertainty in these data. Furthermore, measurements always return integer values rather than real numbers. If temperature, pressure, and humidity are real numbers in the physical world, this digitization introduces additional uncertainty into the representation of the initial conditions the computer uses. **Data assimilation** is the process of transforming actual-world observations into the model state space. This process produces a model of the initial data within the forecasting model state space. Assimilation introduces additional representational uncertainty into the initial data. Scientists work with a model for the total uncertainty representing the error between the measurements assimilated into the forecast model state space, and the ontic or True magnitudes of these quantities believed to be in the actual world. To make a reliable forecast, the model for data uncertainty needs to be both consistent with the observations and consistent with the weather forecasting model.

6.4 Computers, integers and memory

The fact that digital computers can only use integers implies a limit on computer representations when the difference between numbers becomes too small. For instance, computers treat all infinitesimally close real numbers as if they are the same integer. This means there are limits to how well computers can track SDIC. Moreover, all computers have a finite amount of memory no matter how large implying a limit on how many distinct internal states they can have. At some point in any calculation a computer will have to return to a previously used state. Since computers are deterministic devices, they will begin to repeat the same states again, hence the difficulty in numerically confirming aperiodic solutions remain so indefinitely.

6.5 Shadowing

Despite these limitations and challenges computers are an unavoidable tool for scientists' study of complex problems. Indeed, scientists, such as Lorenz, use computers, to study and gain understanding about nonlinear physical systems with great success. Success in this context means that computer model output **shadowed** the time series generated by observing the target system closely enough to be consistent with the model for noise for a sufficiently long enough time.

One lesson of chaos studies is that scientists must pay closer attention to models of noise than when studying linear systems. Given that models of target systems and noise are imperfect, there are limits on the **shadowing time**, how long a computer model's output in state space remains reasonably close to the target system trajectory.

Along with the uncertainty in initial data and the restriction to integer representations, measurements typically contain information on smaller scales than can be modeled in even the most powerful supercomputers. This information gets left out of the initial data as well as the computer model itself. Furthermore, there are variations in the key variables, such as pressure, temperature, and eddies in wind currents for weather, on length scales too small to be represented in any computer model. Since scientists are dealing with chaotic dynamics, there are nonlinearities involved meaning we need appropriate techniques for analyzing models and data. This is where shadowing and **ensemble forecasting** enters.

As contemporary weather forecasting illustrates, chaotic dynamics does not render predictability hopeless. Consider the modeling for hurricane Helene in figure 1.1. Probabilities produced by the model reflect growth in uncertainty from the time the latest observational data is assimilated into the model to the forecasted times and possible locations of hurricane landfall and subsequent movement. Shadowing assumes that Helene's actual trajectory lies somewhere in the model's evolving cone of uncertainty. In other words, the ontic state of the hurricane is located somewhere in this probability distribution.

The key idea of shadowing is this: observations of the weather conditions plus a model of the noise in the observations of the initial conditions are represented in an ensemble. There should be at least one initial state in the ensemble of initial conditions within the uncertainty of the observations forming the initial data assimilated into the computer model such that under iteration the forecasted future states stay close to those of future observations. Hence, if we're using NOAA's observations, the model's time series output tracks closely with or shadows the observations of the evolving storm accurately. A model adequate to our forecasting needs will have at least one such shadowing state, and this state and its evolution will be reflected in the forecast using the following strategy:

1. Collect the initial observations of a weather system and assimilate these into the forecasting model.
2. Vary the initial data for each model run within the uncertainty of the observations producing an ensemble forecast.
3. Compare the probability forecast with the next set of weather system observations.

4. Keep only those model states shadowing the observed behavior of the system.
5. Assimilate the latest observations into the model and repeat steps 2–5.

Such a strategy ensures model accountability to all data up to the present while producing reasonable projections of weather system behavior into the future. Under these assumptions, the strategy produces an ensemble forecast: on 30 out of 100 of the model runs using the ensembles, rain is predicted for Hamburg, Germany tomorrow, while no other outcomes consistently group together. This goes into the weather forecast on the evening news of 30 percent chance for rain in Hamburg tomorrow. Similarly for hurricane forecasting.

The strategy represented by 1–5 only forecasts well if there is a model specifying how to estimate the minimum size of the ensemble needed to be confident it contains a sufficient number of shadowing trajectories. Additionally, we improve forecasting accuracy by reducing observational noise as much as possible. Good (but not perfect) models, statistics, and practices enable NOAA, the European Center for Midrange Weather Forecasting, and other forecasting agencies to produce probability forecasts that shadow weather systems well enough to save many lives and minimize economic damage.

This strategy is improved by making use of stochastic or random noise. Given that computers aren't likely to ever be powerful enough to run forecasting models with resolution down to the smallest scales, scientists inject some random noise into each model run to account for the information the model cannot track (e.g. on scales of tens or hundreds of square meters).[3] Treating the unknowns at smaller length scales as random variables means the computer model for weather forecasting, say, also includes a model for effects that cannot be represented in the computer helping mitigate representational error.

For each model run in strategy 1–5, the ensemble of initial conditions gets a slightly different noise injection amounting to letting the missing information in our data and weather model behave slightly differently on each run. Ensemble weather forecasts using such random noise injections shadow much more accurately than leaving such noise out [5]. Chaos has taught us that noise can be a friend rather than always an enemy.

Despite the modeling issues raised in this chapter, scientists have developed techniques for producing useful forecasts for weather and other systems. For instance, your evening weather forecast is still quite reliable despite the existence of chaotic dynamics: measurements of the initial state of temperature, pressure, humidity, and so forth are taken yielding ensembles of initial conditions. This data is assimilated into the model state space. Scientists can run their forecasting models following steps 1–5 to produce an ensemble forecast. Your weather forecast predicts temperatures that are often very accurate, the probabilities for rain and rain amounts reliable, and so forth. Though there are occasions where the storm system

---

[3] This is apparent randomness (section 1.1) because random number generators depend on deterministic algorithms.

may go slightly further north or south of your town than forecast, meaning you may get somewhat less rain than expected.

Ensemble forecasting techniques are used in weather, climate change, infectious disease spread, and other kinds of modeling, where nonlinear effects can be important. A particularly interesting case is cosmological modeling of galaxy formation and distribution, where chaotic processes at smaller length scales can influence large-scale galaxy evolution (e.g., [6]).

6.6 Machine learning

Techniques such as shadowing, ensemble forecasting, and random noise injection are important advances in modeling nonlinear systems that can exhibit chaotic dynamics. The latest addition to the forecasting toolkit is **machine learning**, sometimes called artificial intelligence. It is a branch of computer science that develops computer algorithms to detect patterns in data, represent and collate data, among other tasks. Machine learning algorithms are designed to minimize a function representing the solution to a problem or a pattern recognition task such as pedestrian recognition. Large data sets are used to train the function to solve the problem (e.g., find the most efficient delivery route) or recognize the pattern (e.g., when people are crossing the street in front of a moving vehicle).

Machine learning algorithms can be trained to recognize chaotic behavior and coupled with weather observations to recognize when a weather system is in a chaotic regime, where nonlinearities are expected to have significant effect [7]. Such algorithms can be used in conjunction with ensemble forecasting models enabling weather forecasters to take chaotic dynamics into account in the construction of their ensemble forecasts. Machine learning algorithms also can make their own forecasts based on these observations [8].

References

[1] Arecchi F T and Meucci R 2008 Chaos in lasers *Scholarpedia* **3** 7066
[2] Bishop R C 2003 On separating prediction from determinism *Erkenntnis* **58** 169–88
[3] Plotnitsky A 2023 'The agency of observation not to be neglected': complementarity, causality and the arrow of events in quantum and quantum-like theories *Phil. Trans. R. Soc.* **381** 20220295
[4] Bishop R C 2002 Deterministic and indeterministic descriptions *Between Chance and Choice: Interdisciplinary Perspectives on Determinism* ed H Atmanspacher and R Bishop (Thorverton: Imprint Academic) 5–31 pp
[5] Palmer T 2024 Stochastic weather and climate models *Nat. Rev. Phys.* **1** 463–71
[6] Genel S *et al* 2019 A quantification of the butterfly effect in cosmological simulations and implications for galaxy scaling relations *Astrophys. J.* **871** 21
[7] Barbosa W A S and Gauthier D J 2022 Learning spatiotemporal chaos using next-generation reservoir computing *Chaos* **32** 093137
[8] Ben-Bouallegue Z *et al* The rise of data-driven weather forecasting. arXiv:2307.10128v2 [physics.ao-ph]

IOP Publishing

An Introduction to Chaotic Dynamics
Classical and quantum
Robert C Bishop

Chapter 7

Quantum influences on macroscopic chaotic systems

With the previous material on chaos in classical (macroscopic) systems in place, we turn to quantum mechanics (QM) and possible effects on macroscopic systems. Is sensitive dependence on initial conditions (SDIC) of classical chaotic systems sufficient to amplify quantum fluctuations such that macroscopic systems are affected? If so, then classical chaotic systems possibly are indeterministic after all. However, there are significant hurdles to the amplification of quantum effects to macroscopic scales.

7.1 SDIC and quantum mechanics

In principle, chaotic dynamics is sensitive to the smallest changes to a system's initial conditions. This sensitivity raises the possibility that macroscopic systems exhibiting chaos might be sensitive to quantum fluctuations. Several authors have argued for this kind of openness of macroscopic chaotic dynamics to quantum influences (e.g., [1–3]).[1]

The central argument supporting QM influencing macroscopic chaotic system—the sensitive dependence (SD) argument—is as follows:

1. For macroscopic systems exhibiting SDIC, trajectories starting in a highly localized region of state space will diverge on-average exponentially fast.
2. Quantum effects limit the precision with which physical systems can be specified to a neighborhood δ in state space of no less than a magnitude of $1/(2\pi/h)^N$, where N is the dimension of the system in question.[2]
3. After sufficient time, two trajectories of the same chaotic system starting in δ will have future states localizable only to a much larger region in state space.

---

[1] For more background and discussion, see [4].
[2] For uncorrelated electrons, the exponent is $2N$.

doi:10.1088/978-0-7503-6453-9ch7

4. Therefore, quantum effects will influence macroscopic systems exhibiting SDIC leading to violations of unique evolution (section 1.3).

This argument captures the intuition that since QM sets a lower bound on the size of δ, if quantum effects are indeterministic, then macroscopic chaotic systems will exhibit indeterminism by amplifying such effects to macroscopic scales.

To illustrate, consider a damped driven pendulum exhibiting chaotic behavior. In principle, if a photon strikes the pendulum, a minute amount of momentum from the photon can be transferred to the pendulum and amplified by SDIC affecting the behavior of a macroscopic system.

One way to support the SD argument has been by pointing to strange attractors (section 4.1). For instance, one might argue that their self-similar structure on all scales implies that any trajectories either (i) approach closer and closer to the quantum level increasing prospects for QM to affect the chaotic pendulum, or (ii) if there is some length scale below which self-similar structure doesn't apply, then the trajectories have reached the quantum domain [2], p 157. However, as noted above, actual-world systems likely only harbor pre-fractals; strange attractors are a mathematical feature of state space, not the physical space of macroscopic systems such as damped driven pendulums. Recall one of the faithful model framework's assumptions is that the model state space represents all the possibilities of the physical system (section 1.9). In this case, the state space overestimates the physical possibilities of the actual-world system. The dynamics of a chaotic pendulum represents the ongoing stretching and folding of trajectories, not the possibility that the pendulum dynamics somehow approaches the microscopic realm of QM. There is no interface with quantum effects to be had.

Moreover, the SD argument suffers from a confusion of state spaces. The argument presupposes that the same state space encodes the possibilities for both classical as well as quantum systems, but this isn't the case. State spaces characterizing classical mechanical systems, such as pendulums, are qualitatively distinct from state spaces characterizing quantum systems [5].

These technical caveats notwithstanding, can the photon's minute momentum contribution be amplified by the chaotic dynamics affecting the pendulum's behavior? After all, this addition to the momentum of the chaotic pendulum's arm is different from what the momentum would have been in the absence of the photon's interaction.[3] These represent slight differences in initial conditions and SDIC potentially does the work.

However, there are important constraints that must be accounted for before any amplification of the momentum perturbation could affect the macroscopic pendulum.

---

[3] The electromagnetic interaction between the photon and pendulum is complex. A relatively simpler interaction would be sending silver atoms through a Stern–Gerlach apparatus, where an atom of either spin up or spin down strikes the pendulum's arm. The analysis remains unchanged, however.

7.1.1 Damping constraints

First, although the mathematical model for our damped driven pendulum represents friction as a constant, friction processes generally aren't constant but involve a number of microscopic processes [4], p 538–41. Roughly, given that friction involves molecular bonds between surfaces, the momentum imparted by the photon to the chaotic pendulum must cause some bonds to form (break) that wouldn't normally form (break) on a time scale due purely to the macroscopic mechanical process of two surfaces sliding across one another. It is the case that models of frictional vibration due to stick-slip motion can exhibit chaotic behavior [6]. Hence, suppose the smallest possible difference delaying the formation or breaking of one bond could make a difference relevant for SDIC. According to the SD argument, this is sufficient to influence the chaotic pendulum.

Valence electrons move about like a gas producing the forces responsible for molecular bonds between two metal surfaces.[4] Molecular bonds due to interatomic potentials typically have strengths on the order of 10^{-13} g cm s^{-2}, placing a lower bound on the magnitude quantum effects need to break such a bond for dry surfaces.[5] The upshot for the photon+chaotic pendulum system is that if the photon's perturbation is amplified to a sufficient magnitude on a short enough time scale, it could affect an intermolecular bond. Interatomic potential strengths present a tight, but not impossible constraint for the photon perturbation to affect the chaotic pendulum's behavior.[6]

7.1.2 Nonlinear dynamics constraints

Another constraint on the SD argument is its reliance upon global Lyapunov exponents. As noted in section 5.4, global Lyapunov exponents only measure infinitesimal growth in uncertainties (Appendix C). So long as the amplification rate of the photon's contribution to the chaotic pendulum's momentum remains infinitesimal, the pendulum's behavior will never be affected.

This means that the SD argument must invoke finite-time, local Lyapunov exponents to succeed. For the sake of argument, assume the perturbation is of sufficient strength to be amplified to a magnitude sufficient to break a molecular bond perhaps on a time scale shorter than mechanical action of the pendulum alone would break the bond. There still are several possible outcomes for the photon's contribution. Given that for local Lyapunov exponents the growth in uncertainties is characterized by the local point-by-point dynamics the uncertainty in momentum generated by the photon may (1) converge rather than grow, (2) grow slower than necessary to break a bond on the appropriate time scale, or (3) grow rapidly enough to break the bond before mechanical action does so. The first two possibilities

---

[4] Processes such as electron tunneling [7], electron–hole pair production [8, 9]), and dipole field fluctuations [10] could also be relevant for mechanical friction due to chaotic dynamics.

[5] Metal surfaces with no contaminants such as oxides or lubricants.

[6] One might argue quantum perturbations cancel out to zero, but experimental evidence indicates otherwise (e.g., [11]).

preclude the photon having any effect on the chaotic pendulum's evolution in which event the SD argument would fail.

7.2 Chaos and indeterminism in macroscopic systems I

The question whether quantum effects can render chaotic macroscopic systems indeterministic still remains. Recall, that determinism is one of the necessary conditions for a dynamical system to exhibit chaotic dynamics (section 1.4 and 5.1). For determinism to fail in a dynamical system, it must be the case that starting the system with identical initial conditions leads to different behavior (section 1.3). If a chaotic macroscopic system can amplify a small quantum perturbation, would this render the system indeterministic?

The answer to this question is subtle. In QM, wave functions contain all the information about a given quantum system. For the identical initial preparation of a quantum state, wave functions undergo identical Schrödinger evolution until interacting with some system (e.g., a measurement apparatus).

Return to the example of the photon+chaotic pendulum system. A reasonable assumption is that the moment the chaotic pendulum registers the photon's impact an irreversible act of momentum transfer to the pendulum has occurred. Under versions of QM in which measurement processes don't indeterministically collapse wave functions (e.g., where there is only Schrödinger evolution or under David Bohm's theory), there is no quantum indeterminism for SDIC to amplify, only deterministic noise. Under other versions of QM, where measurement processes indeterministically collapse wave functions to one sharp value from the super-position of possible values, say, there would be indeterminism for SDIC to amplify.

But this isn't the full story as there is an ambiguity in what counts as the initial state of the photon+chaotic pendulum system and, therefore, the implications for indeterministic versions of QM. As John Bell noted,

> The problem is this: quantum mechanics is fundamentally about 'obser-vations'. It necessarily divides the world into two parts, a part which is observed, and a part which does the observing. The results depend in detail on just how this division is made, but no definite prescription for it is given. All we have is a recipe which because of practical human limitations is sufficiently unambiguous for practical purposes [12], p 124.

There are different ways an observer-observed distinction might be drawn. In the case of the photon+chaotic pendulum system, one option is specifying the system before the interaction takes place as the initial state. Identical initial preparation of the pendulum and photon states leads to the same initial conditions.[7] The system is indeterministic under indeterministic collapse of the photon wave function resulting from the measurement-like interaction between photon and pendulum.

---

[7] Photons are initially in the same pure state representing a superposition of possible energy and frequency eigenstates. This wave function as a whole plus the state of the pendulum are the initial conditions.

A second option is specifying the moment of interaction as the initial state resulting in different initial conditions every time the experiment is run due to the indeterministic collapse of the wave function. Although photons may be prepared in the same initial state for each experimental run, indeterministic wave function collapse means the photon imparts a different magnitude of momentum to the chaotic pendulum.[8] However, this implies that there are different initial conditions on each experimental run leading to different pendulum behaviors consistent with unique evolution. Determinism would be preserved for the chaotic pendulum amplifying the momentum perturbation. Under such circumstances, the chaotic pendulum would exhibit apparent randomness (section 1.1) because of lack of empirical access to the precise amount of momentum the photon imparts.

For this simplified case, unique evolution is violated if the photon+chaotic pendulum system is treated as the initial state when there is Schrödinger evolution and indeterministic wave function collapse. Whether the photon's perturbation leads to indeterminism in the macroscopic system depends on the version of QM, the solution to the measurement problem (e.g., [13]) and, in concrete cases, what counts as the initial condition. Furthermore, constraints on the amplification of the perturbation due to damping must be overcome, which depends on the local finite-time dynamics during the photon's interaction with the pendulum.

7.3 Chaos and indeterminism in macroscopic systems II

Recent work on stochasticity in turbulence indicates that fluctuations at molecular scales can randomize large-scale fluid flow models at high Reynolds number leading to universal statistics [14]. Three cases can be distinguished for these effects in fluid flow models. The first case is the presence of positive global Lyapunov exponents in fluid flow models for high Reynolds number [15]. These models are subject to all the limitations associated with such exponents (section 5.3 and 5.4, appendix C). An examination of the finite-time local Lyapunov exponents is the only way to draw relevant inferences about whether chaos for such fluid models can amplify molecular fluctuations.

The second case is when Lipschitz continuity breaks down [16]. Lipschitz continuity is a stability condition for determinism [17]. When it fails only probability distributions can be defined for the set of solution trajectories. The probability distribution is nondegenerate meaning that trajectories can branch from the same spatial point given the same initial conditions. Hence, the model behavior is intrinsically stochastic and solutions of model equations will not be unique. The failure of unique evolution implies that chaotic behavior isn't possible for these latter models and any sensitivity to small changes is due to the failure of the principle of linear superposition in nonlinear models.

A third case was studied by Lorenz [18], where sensitive dependence due to nonlinearity in fluid flow models can lead to a breakdown in continuous dependence

---

[8] Since there is a continuum of possible position and momentum states for the photon that could be actualized in interaction with the pendulum, the probability of achieving the same momentum perturbation on multiple runs is of measure zero.

on initial conditions (e.g., [19]). For the model he derived, Lorenz assumed that the observations of initial conditions assimilated into such models could be made arbitrarily small.[9] He demonstrated numerically that any uncertainty in initial conditions at the smallest length scales that can be modeled on a computer produce stringent finite predictability limits even if the uncertainty can be reduced arbitrarily. This means that there is a finite time limit after which the deterministic model's forecasts would fail, illustrating the effects of apparent randomness: 'two states of the system differing initially by a small observational error will evolve into two states differing as greatly as randomly chosen states of the system within a finite-time interval, which cannot be lengthened by reducing the amplitude of the initial error' [18], p 289. The culprit is nonlinearity combined with multiple length scales. No Lyapunov exponents, no chaotic dynamics; only nonlinear dynamics across multiple length scales. Lorenz compares this situation observationally to irreducibly random systems where indeterminism cannot be removed. Such systems 'cannot be perfectly predicted even when the uncertainty of the initial state is reduced to zero' [18], p 304. He emphasizes that no appeals to quantum indeterminism need be made for this comparison. Nonetheless, it's important to realize that unpredictability itself doesn't imply a failure of determinism [20].

Hadamard argued a breakdown in continuous dependence on initial conditions was a sign of a mathematically ill-posed problem [21], p 138–42. Following his lead, many have thought that such discontinuous dependence on initial conditions, while manifesting in some mathematical models, were physically meaningless, though Lorenz [18] argues the opposite. Recently, Simon Thalabard *et al* [22] demonstrated that switching from a framework focusing on trajectories to one of probability distributions can transform some of these mathematical models from ill-posed to well-posed while still maintaining the stringent forecasting limits Lorenz studied. Their work demonstrates that despite predictability limits the Kelvin–Helmholtz instability [23] possesses well-defined statistical properties at finite times though these properties are not sensitive to the nature of small-scale details of the fluid model. This instability is relevant to the growth of a shear fluid layer from an initially discontinuous velocity profile. The probabilistic model involves infinitesimal perturbations in the fluid viscosity and energy at the smallest scales that are amplified up to finite effects in finite time indicating super-exponential growth in uncertainties. Indeed, the apparent randomness of the solutions appears in finite time. The evolution of the probability distributions leads to universal statistical properties independent of the micro details of the perturbations in the limit of large times.

This third case of fluid flow models indicates limits to the applicability of determinism for such models for finite time ranges. This failure is consistent with mathematical results that uniqueness of differential equation solutions typically is only guaranteed for finite times (e.g., [24]). Given that these numerical results depend only on vanishingly small perturbations at the smallest scales, they are

---

[9] Lorenz didn't believe this was possible, but makes the assumption for the sake of the argument. We saw earlier why this assumption is problematic (section 6.2).

suggestive that actual-world fluid systems potentially could be sensitive to molecular scale or smaller fluctuations potentially serving as a route for quantum effects to influence fluid systems. Yet, we currently have no evidence that these kinds of spontaneous stochastic phenomena occur in actual fluids [14, 22]. Additionally, there is the question of the faithful model framework (section 1.9) regarding the relationship of these models—or at least some features of them—to actual-world systems. It should be emphasized that although predictability of individual trajectory solutions for such models may be limited, a wealth of statistical predictions for such system are possible. This is an ideal situation for ensemble and shadowing forecasting techniques (section 6.5).

References

[1] Barone S R *et al* 1993 Newtonian chaos + Heisenberg uncertainty = macroscopic indeterminacy *Am. J. Phys.* **61** 423–7

[2] Hobbs J 1991 Chaos and indeterminism *Can. J. Phil.* **21** 141–64

[3] Kellert S H 1993 *In the Wake of Chaos* (Chicago, IL: University of Chicago Press)

[4] Bishop R C 2008 What could be worse than the butterfly effect? *Can. J. Phil.* **38** 519–48

[5] Bishop R C 2024 *The Physics of Emergence* 2nd ed (Bristol: IOP Publishing) https://doi.org/10.1088/978-0-7503-6367-9

[6] Bengisu M T and Akay A 1992 Interaction and stability of friction and vibrations *Fundamentals of Friction: Macroscopic and Microscopic Processes vol 220 of NATO ASI Series* ed I L Singer and H M Pollock (Dordrecht: Kleuwer Academic) 553–66 pp

[7] Persson B N J and Demuth J E 1985 Determination of the frequency-dependent resistivity of ultrathin metallic films on Si(111) *Phys. Rev.* B **31** 1856–62

[8] Persson B N J and Zaremba E 1985 Electron-hole pair production at metal surfaces *Phys. Rev.* B **31** 1863–72

[9] Persson B N J 1991 Surface resistivity and vibrational damping in adsorbed layers *Phys. Rev.* B **44** 3277–96

[10] Persson B N J and Ryberg R 1981 Vibrational interaction between molecules adsorbed on a metal surface: the dipole-dipole interaction *Phys. Rev.* B **24** 6954–70

[11] Mak C *et al* 1994 Atomic-scale friction measurements on silver and chemisorbed oxygen surfaces *Thin Solid Films* **253** 190–3

[12] Bell J S 1987 *Speakable and Unspeakable in Quantum Mechanics* (Cambridge: Cambridge University Press) https://doi.org/10.1017/CBO9780511815676

[13] Quantum Theory and Measurement ed J A Wheeler and W H Zurek 1983 *Quantum Theory and Measurement* (Princeton, NJ: Princeton Univversity Press) https://doi.org/10.1515/9781400854554

[14] Bandak D 2024 Spontaneous stochasticity amplifies even thermal noise to the largest scales of turbulence in a few eddy turnover times *Phys. Rev. Lett.* **132** 104002

[15] Bernard D *et al* 1998 Slow modes in passive advection *J. Stat. Phys.* **90** 519–69

[16] Eijnden W E and Eijnden E V 2000 Generalized flows, intrinsic stochasticity, and turbulent transport *Proc. Natl Acad. Sci.* **97** 8200–5

[17] Bishop R C and beim Graben P 2016 Contextual emergence of deterministic and stochastic descriptions *From Chemistry to Consciousness: The Legacy of Hans Primas* ed H Atmanspacher and U Müller-Herold (Cham: Springer) 95–110 pp

[18] Lorenz E N 1969 The predictability of a flow which possesses many scales of motion *Tellus* **21** 289–307

[19] Palmer T N *et al* 2014 The real butterfly effect *Nonlinearity* **27** R123

[20] Bishop R C 2003 On separating prediction from determinism *Erkenntnis* **58** 169–88

[21] Hadamard J 1922 *Lectures on Cauch's Problem in Linear Partial Differential equations* (New Haven, CT: Yale University Press)

[22] Thalabard S *et al* 2020 From the butterfly effect to spontaneous stochasticity in singular shear flows *Commun. Phys.* **3** 122

[23] Matsuoka C 2014 Kelvin-Helmholtz instability and roll-up *Scholarpedia* **9** 11821

[24] Arnold V I 1988 *Geometrical Methods in the Theory of Ordinary Differential equations.* (New York: Springer) https://doi.org/10.1007/978-1-4612-1037-5

IOP Publishing

An Introduction to Chaotic Dynamics
Classical and quantum
Robert C Bishop

Chapter 8

Quantum mechanics and quantum chaos

Having discussed the possibilities for quantum fluctuations to influence macroscopic chaotic systems, we turn to quantum mechanics (QM) proper. What is quantum chaos and how does it differ from classical chaos? Our definitions become crucial here. Surprising connections between quantum semiclassical, and classical systems arise, raising questions about how to understand the correspondence principle and the relationship between the microscopic phenomena of QM and the macroscopic world.

8.1 Defining quantum chaos

Quantum chaos largely is the study of the relationship between chaotic dynamics in macroscopic models and their quantum counterparts. Some, such as Sir Michael Berry, have argued these studies should be called **quantum chaology** [1, 2]. Physics journals haven't accepted this terminology. While quantum chaos is qualitatively distinct behavior from classical chaos (see below and chapter 9), the same can be said for the differences between QM and classical mechanics. Even though some confusion might be avoided using quantum chaology terminology, the parallelism between QM and classical mechanics terminologies has been sustained.

As might be expected from the classical case (chapter 5), defining quantum chaos as parallel to classical chaos is problematic. The presence of some form of stretching and folding mechanism associated with nonlinearity appears to be a necessary condition for classical chaos (section 1.7). However, Schrödinger's equation is a linear evolution equation implying QM is a linear theory: quantum states starting out initially close in Hilbert space norm remain close throughout their evolution. It is possible for wave packets to bifurcate and/or merge, but the dynamics of wave packets is distinctly different from that of macroscopic system trajectories.

Determinism is another necessary condition for classical chaos (section 1.3). While there are some deterministic alternatives, QM is considered to be an archetypical example of an indeterministic theory. Questions can be raised about

doi:10.1088/978-0-7503-6453-9ch8

the status of indeterminism in QM, but there is good evidence quantum systems fail to exhibit unique evolution.

Setting aside how problematic global Lyapunov exponents are, they cannot be defined in Hilbert space.[1] Moreover, wave functions must have finite sizes due to Heisenberg uncertainty, so infinitesimal separations cannot be defined and expectation values will disagree with the state space averages calculated for quasi-classical trajectories as in semiclassical systems. This circumstance also implies aperiodic trajectories (section 1.6) cannot exist in Hilbert space, meaning every candidate for a necessary condition for chaos is absent in QM. Furthermore, because surfaces in Hilbert space aren't well defined for quantum systems, Poincaré surface-of-section plots aren't well defined. Importantly, while classical state spaces support fractal structure (section 4.1), Hilbert spaces cannot. The important differences between quantum and classical chaos are due to differences between Hilbert space and state spaces for classical mechanics.

8.2 Semiclassical systems

For these reasons, quantum chaos studies the behavior of quantum systems and their semiclassical counterparts. Pioneering work on semiclassical systems was carried out by John Van Vleck [3] (1899–1980) though Martin Gutzwiller (1925–2014) also made seminal contributions (see [4], and references therein) among others.

It is important to clarify that semiclassical modes don't represent **mesoscopic physics**, the physics of systems somewhere between quantum and macroscopic length and time scales that often are treated by renormalization group methods [5–7]. Instead, the so-called semiclassical approximation aims to address the relationship between microscopic (quantum) physics, described by Schrödinger's equation, and macroscopic (classical) physics, described by the Hamilton–Jacobi equation.

One way to construct semiclassical models is to start with a classical chaotic system and quantize it by replacing observables in the equations of motion with their corresponding quantum operators (e.g., position and momentum operators replace the classical position and momentum observables, as in $\frac{\partial S}{\partial q_k}$ is replaced by $ih\frac{\partial}{\partial q_k}$, where S is the classical action, and q_k represents position). Such procedures introduce Planck's constant into the semiclassical model. Alternatively, one can start with a quantum model and 'let Planck's constant go to zero' to produce a 'classical' version.[2]

There are a number of remarkable behaviors exhibited by quantized semiclassical systems that are interesting in their own right. How chaotic dynamics in these systems is related to the quantum domain and the validity of the correspondence principle (section 9.1) are questions that we will turn to in due course. Suffice it to

---

[1] A kind of generalized Lyapunov exponent can be computed describing the asymptotic growth rate of the norm of a product of random matrices acting on a vector. Both global (as the number N of products grows to infinity) and finite-N generalized exponents can be calculated. But these have no relationship to the global Lyapunov exponents of classical chaotic dynamics.

[2] Van Vleck [3] was an early and perhaps the most rigorous pioneer of these techniques.

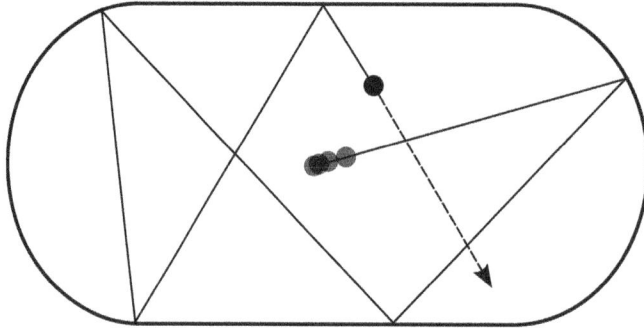

Figure 8.1. An example of a stadium billiard. Courtesy of Wikimedia Commons.

say, these studies reveal evidence that the relationship between the quantum and classical domains is more subtle than textbook claims (chapter 9).

Quantum chaos studies usually involve quantum systems describable by a finite number of parameters or finite amount of information. Such studies often focus on universal statistical properties independent of the quantum systems under investigation. The kinds of statistical properties studied include statistics of energy levels and semiclassical structures of wave functions. These statistical properties are relevant for quantum state transitions, ionization, and quantum phenomena found in atomic and nuclear physics, solid state physics, and even quantum information. Prototypical systems studied are quantum billiards, quantum kicked rotors, single periodically driven spins, and coupled spins.

Quantum billiards are workhorse models for quantum chaos. Classical billiards are two-dimensional systems where point particles bounce elastically off edges. Figure 8.1 is an example of the stadium billiard. The angle of reflection from the boundary equals the angle of incidence, and the dynamics is defined by the shape of the boundary.

Numerous analytic results exist for classical billiards. A chaotic billiard is where parameter magnitudes lead to chaotic behavior of the balls (e.g., aperiodic orbits and SDIC). One can produce quantum billiards by using Schrödinger's equation to describe wave functions reflecting off boundaries (requiring the potential equal zero inside the boundary but be infinity on the boundary). Alternatively, one can start with the equations describing a classical billiard and quantize the position and momentum observables to obtain semiclassical quantized billiards.

8.3 Isolated quantum systems

Start with the simpler case of isolated quantum systems. An immediate contrast with classical system is that bounded isolated quantum systems have discrete energy spectra while classical bounded systems have continuous energy spectra. In dynamical systems theory, discrete spectra are associated with integrable systems where chaotic behavior is impossible.

There are other significant differences between isolated bounded classical and quantum systems. Berry *et al* [8] demonstrated that semiclassical quantum systems can mirror behaviors of their corresponding classical systems only up to the Ehrenfest time t_E, of the order $\ln(q)/\hbar$ seconds, where q corresponds to a relevant quantum number and $\hbar$ is Planck's constant divided by 2π. The log time reflects the exponential instability of classical chaotic trajectories. Since semiclassical studies treat Planck's constant as a parameter, where reducing its magnitude supposedly leads to the classical domain, the smaller the value, the more 'classical' a system's behavior. As $\hbar \rightarrow 0$, t_E grows.

On the other hand, t_E is supposed to be a measure of when quantum wave packets have spread too much to mimic classical trajectories and the Ehrenfest theorem breaks down. Gaussian wave packets centered on classical trajectories are thought to be able to mimic such trajectories up to t_E before becoming too spread out over the energy surface. Therefore, there are two important effects at work in these semiclassical systems over time, the coalescing of classical chaotic trajectories, and the spreading of quantum wave packets. Combined with the linearity of Schrödinger's equation, this is further evidence that properties of classical chaos don't exist for bounded isolated quantum systems.

Although t_E is an important limit on quantum wave packets tracking classical trajectories, there are interesting behaviors of semiclassical models with classical chaotic counterparts on longer time scales. Steven Tomsovic and Eric Heller [9] demonstrated that full quantum solutions compared with suitably chosen semi-classical solutions for some billiards problems exhibit excellent agreement well after t_E, including fine details of energy spectra. Their semiclassical models remain accurate for up to a time that scales with $1/\sqrt{(\hbar)}$.

Typically, however, limitations on quantum systems exhibiting the marks of classical chaos for bounded isolated systems are due to the time scale at which quantum evolution becomes dominated by fluctuations. This time scale is given by the Heisenberg time, $t_h = 2h\rho$, where $\rho = dE_n/dn$ is the density of states. For the isolated quantum systems discussed in this section, t_h is extremely short, offering no opportunity for classical chaotic behavior before it would be swamped out by fluctuations.

8.3.1 Quantized chaotic system behavior

Various results demonstrate that strongly ergodic (Appendix A) classical billiards, when quantized, exhibit quantum ergodicity. Nonetheless, this isn't the same as showing that a quantized classical chaotic system exhibits the marks of classical chaotic behavior. There are no examples of the latter due to the reasons discussed earlier in this chapter.

These differences can be illustrated by stadium billiards. Figure 8.2 shows a chaotic stadium billiard where the trajectories fill the space (left image). In contrast, quantum stadium billiards produce scarring patterns that aren't space filling. Recently, quantum chaotic billiard scaring was experimentally confirmed (Figure 8.2, right; [10]).

Nevertheless, there are interesting numerical results on quantum interference in quantized classical billiards (e.g., [11]). Consider a double slit with a width of three de Broglie wavelengths. Enclose the photon source in a two-dimensional wave

Classical Chaos   **Quantum Chaos**

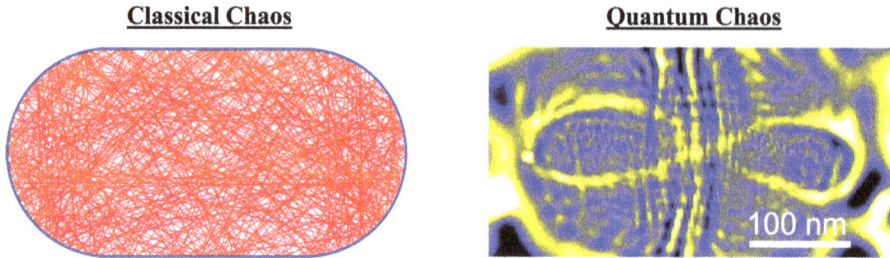

Figure 8.2. Classical chaotic billiard trajectories are space filling (left). Quantum chaotic billiards produce scaring, which has been experimentally confirmed (right). Courtesy of the Velasco Lab.

resonator with the shape of a classical billiard, producing wave packets sharply peaked in momentum with a spatial spread the width of the resonator. Select Gaussian wave packets of initial average energy one 1600th of an excited state of the quantized billiard and send them toward the resonator's double-slit opening. If the resonator corresponds to a classical chaotic billiard, there is almost no quantum interference.[3] On the other hand, if the shape of the resonator corresponds to a classical regular billiard, then the expected interference patterns emerge. Hence, whether the corresponding classical billiard is chaotic or not determines whether the quantum analog exhibits interference. Moreover, the presence of periodic orbits in the analog classical systems largely determine the properties of semiclassical systems [12].

Typically, billiards are modeled using point particles but more physically realistic systems—particles with nonzero radius—may be better suited for studying quantum chaos (e.g., [13]). Transitioning from a point particle to one of finite radius is equivalent to making the billiard table smaller by a factor related to the size of the particle radius r. This amounts to a scale transformation. For a classical billiard system with point particles exhibiting regular dynamics, transforming from $r = 0$ to $r > 0$ can lead to chaotic behavior even though parameter magnitudes in the equations describing the dynamical system remain unchanged. The reverse transition also is possible: a chaotic classical billiard with finite-radius particles transforming to point particles exhibits regular orbits with no changes to other parameter magnitudes. In effect, r functions as an additional parameter enabling chaotic behavior in classical billiards.

Finite-radius particle billiards may be a more suitable context for studying quantum chaos in the semiclassical limit because wave functions are always of finite size in quantum systems. Instead of calculating Lyapunov exponents, an out-of-time-ordered correlator (OTOC) can be defined for quantum billiard systems among others [14]. OTOCs can exhibit exponential growth, which some have taken to be analogous to the global Lyapunov exponent in the semiclassical limit $h \to 0$ (e.g., [15]).[4] These correlators quantify the degree of noncommutativity in time between two operators. While OTOCs exhibit exponential growth for some

---

[3] In the classical case, the multiply reflected waves become randomized in phase.
[4] OTOCs were introduced to measure the instabilities in electron trajectories scattered by impurities in condensed matter physics [16].

semiclassical billiards, the analogy with global Lyapunov exponents breaks down for other quantum systems (e.g., [17]). Moreover, some classical billiards exhibiting regular behavior have semiclassical analogs exhibiting exponential growth of OTOCs [15, 18, 19], undermining analogies between classical and quantum chaos. It is possible to show that in the limit as $h \to 0$, OTOCs can have an exponential dependence parameterized by global Lyapunov exponents when the corresponding classical dynamics is chaotic [14]. For the more general case of Heisenberg operators, it has been demonstrated that OTOCs show no exponential instability resulting in zero Kolmogorov–Sinai entropy [20].

8.3.2 The quantum chaos conjecture

The relationship between classical chaos and quantum chaos turns out to be more delicate than is often portrayed in the literature. The classical-to-quantum direction often uses the Gutzwiller semiclassical trace formula in the limit $h \to 0$ ([21]; see also [4], chapter 17).[5] The quantum-to-classical direction is much more difficult and fraught with conceptual problems (chapter 9). Standard approaches, here, begin with a quantum analog to a classical chaotic system and then derive a semiclassical system representing the quantum system in some kind of classical limit [1, 22]. This work results in statistics of suitably normalized energy levels for the semiclassical systems with universal features. For classical systems that behave non-chaotically, the energy levels of the semiclassical systems approximate a Poisson distribution where small level spacings dominate.

In contrast, when classical systems behave chaotically, the energy levels of the corresponding semiclassical systems take on a distribution originally derived by Eugene Wigner [23] to describe nuclear energy spectra.[6] These latter distributions depend only on symmetry properties (e.g., the presence or absence of time-reversal symmetry). Originally derived for heavy nuclei, a key assumption is that, in the Heisenberg representation using typical basis vectors, the matrix elements of the Hamiltonian can be treated as if they are Gaussian random numbers. This produces a model, known as a **random matrix model**, having no free parameters and is invariant under a wide range of change of bases [23, 25, 26].[7]

Many simple QM systems that have semiclassical analogues, where the corresponding classical models exhibit both chaotic dynamics and universal energy level fluctuations, are well described by Wigner's methods. These are low-dimensional systems, such as billiards, having few degrees of freedom and no interactions taking place at different spatial scales. Work on these systems has led to the **quantum chaos conjecture**:

---

[5] It is worth noting that Gutzwiller calls all semiclassical models 'classical' in [4] because the equations can all be solved using techniques from classical mechanics.

[6] For discussion, see [24].

[7] Wigner's random matrix model is very successful at describing energy spectra of a variety of complex quantum systems [26–29].

Definition 8.1. Quantum chaos conjecture: short-range correlations in the energy spectra of semiclassical quantum systems, which are strongly chaotic in a classical limit, obey universal fluctuation laws based on ensembles of random matrices without free parameters.

Known as the Bohigas, Giannoni, Schmit conjecture [28], it proposes a relationship between classical chaos, random matrix theory, and semiclassical analogues of isolated quantum systems.

An important motivation for the quantum chaos conjecture is evidence accumulated over the decades that energy spectra of simple non-integrable semiclassical models derived from classically chaotic models contain fluctuations falling into a universality class described by random matrix theory. Ian Percival [30] first proposed that the statistics of random matrices could be used to formally distinguish the behavior of semiclassical quantized classically chaotic systems from the behavior of semiclassical quantized integrable systems lacking chaotic dynamics.[8] This requires an assumption that classical dynamics is a limit of quantum dynamics through the correspondence principle. If so, the kind of behavior exhibited by the classical dynamics related in this way should be reflected in the corresponding spectra.[9]

Berry and Michael Tabor [31] demonstrated that integrable semiclassical systems of two or more degrees of freedom exhibit the same statistical properties as a pure random process.[10] Energy levels are arranged with statistical independence as if one was drawing marbles out of an urn. Along with the original computer simulations of Oriol Bohigas *et al* [28], other numerical and theoretical support for the conjecture has been provided [32–35]. Some of these results are only valid for times up to t_E. Nonetheless, the conjecture remains to be proven.

A standard example of the conjecture is spectra of semiclassical billiards corresponding to classical billiards exhibiting chaos characterized by a Gaussian ensemble of real random matrices. A hydrogen atom in a strong magnetic field is characterized by Gaussian ensembles of real random matrices as is the excitation spectrum of nitrogen dioxide molecules. Hence, there is theoretical and experimental support for an inference to the best explanation supporting the quantum chaos conjecture.

These classical results are evidence that seeking analog results for quantum systems having a relationship to classical chaotic systems is reasonable. The quantum chaos conjecture implies the energy spectra for the semiclassical analogs of classical chaotic systems are structurally similar as those in the corresponding classical systems. The conjecture focuses on semiclassical systems, implying the structure of the energy spectra of such systems is dependent on chaos in the corresponding classical systems not on any classically chaotic behavior in quantum or semiclassical systems. Nevertheless, note that there are many regular systems

---

[8] Though he uses the language of 'random distributions' rather than random matrices.
[9] We will explore reasons to doubt this simple relationship in chapter 9.
[10] The exception being pathological cases, such as the harmonic oscillator.

whose energy and other statistics are described by random matrices. Therefore, the fact that any given system, quantum, semiclassical, or classical, has characteristics described by random matrices isn't sufficient to distinguish classical chaos though some of the quantum chaos literature suffers from this confusion (e.g., [36]). Similarly, problems arise for treating thermalization results from chaotic dynamics in classical statistical mechanics and thermalization in quantum systems as distinctive marks of classical chaos (e.g., [37]). There are numerous ways systems can approach thermodynamic equilibrium, meaning thermalization isn't a unique diagnostic of chaotic dynamics.

For instance, most work relating thermalization in quantum systems to classical chaos depend on global Lyapunov exponents or analogs in QM that technically are disanalogous to classical chaos. An interesting example is [38]. The authors explore the relationship between the behavior of a quantum infinite spin chain system and its dynamics projected on to variational states in an effective classical Hamiltonian dynamics. The classical dynamics is chaotic, which leads to thermalization and positive global Lyapunov exponents. Meanwhile, OTOCs are used as the analog to Lyapunov exponents in the quantum spin chain. One can track the thermalization over time in the classical model with entanglement growth in the quantum spin chain and even compare the Kolmogorov–Sinai entropy (Appendix A) of the classical dynamics with that of the entangled quantum spin chain states. However, none of the QM properties represent marks of classical chaotic dynamics. Chaos only occurs in the classical model; the link between the quantum and classical dynamics is provided by a quantum-to-classical projection on to classical states.

As discussed in chapter 5, definitions such as $chaos_{Dexp}$ are needed for classical chaos. By themselves, neither statistical signatures, such as random matrices, nor thermalization uniquely distinguish chaotic dynamics.

8.3.3 Quantum-to-classical questions[11]

There are serious questions about these quantum-to-classical studies that often go unaddressed, however.[12] As indicated above, semiclassical systems are derived using various procedures, but these procedures don't yield actual classical systems that are assumed to be limiting cases of quantum systems. The limiting relations between quantum and classical domains are significantly different than semiclassical approaches (chapter 9). One reasons the quantum chaos conjecture remains unproven is that inappropriate notions of 'classical limit' are invoked. The energy level statistics for quantum billiards in the semiclassical counterparts to classical billiard systems may share universal properties. Nonetheless, the actual behavior of the trajectories in classical and quantum systems is substantially different.

As noted earlier, quantum systems are linear while classical chaos is a result of nonlinearities. Quantum chaos studies typically concentrate on scattering processes, such as quantum billiards, and externally driven quantum systems (section 8.4),

---

[11] For a philosophically-oriented discussion of these issues see
[12] For more discussion see chapter 9.

focusing on time evolution and energy level structure. Predictability limits are a feature of classical chaotic systems. However, quantum unpredictability is due to other sources than classical chaos, such as commuting observables undergoing complicated dynamics.

Some quantum systems exhibit bifurcation behavior. Under some circumstances rotating molecules can undergo several consecutive qualitative changes interpreted as bifurcations, for example [39, 40]. Currently it's unknown whether a series of bifurcations in such systems could eventually lead to a transition to some form of classical chaotic behavior. We have seen some reasons to doubt such transitions can occur. Additionally, qualitative behaviors exhibited by quantum bifurcations never lead to a series of limit cycles as observed in classical chaotic models such as the logistic map. Instead, energy levels get reorganized among energy bands in isolated quantum systems [41]. Or there is a transition to a superposition of states. Meanwhile, the corresponding classical chaotic systems choose one or the other branch of a pitchfork bifurcation as in figure 3.8.

Quantum chaos in isolated systems has produced results having interesting relationships with integrable and non-integrable classical systems. Moreover, there are some notable experimental statistical results (e.g., [42–44]). The limitation to isolated quantum systems means that the state spaces of these systems generally don't allow the formation of dynamical structures typically associated with classical chaotic systems. Consider an example discussed in [45], a quantum Hamiltonian operator for an N-dimensional torus: $\frac{1}{2}(g_k n_k + n_k g_k)$, where $n_k = -i\partial/\partial\theta_k$ and θ_k is an angle variable, and $d\theta_i/dt = g_i(\theta_k)$ for $i, k = 1, 2, 3, \ldots, N$. The momentum probability density grows exponentially fast in this system. Still, this isn't analogous to the growth in uncertainties exhibited by classical chaotic systems.[13]

It is time to turn to interacting quantum systems to assess the prospects for chaos.

8.4 Interacting systems

The numerical results of the double-slit/billiard wave resonator described in section 8.3.1, suggest a possibility for applying quantum chaology to the quantum measurement problem. Quantum measurement models describe the destruction of coherent quantum states as an effect of external noise, such as the environment, or interaction with a detector. Studying quantum chaos may aid the development of a dynamical theory of quantum decoherence due to the interaction between classical chaotic, or non-integrable, macroscopic systems and coherent quantum states producing the incoherent mixtures observed in measurement devices. Measurements are quintessential examples of interacting systems.

The same statistical behavior as captured by random matrix models often occurs in interacting quantum systems and semiclassical counterparts. Hence, for suitable parameter magnitudes, quantum chaos exists in open quantum systems. Some have analyzed driven quantum many-body systems using combinations of

---

[13] Recall that there are problems with looking to exponential growth itself as a distinguishing mark of classical chaos (section 5.3).

operators, such as the dual transfer matrix product, and found what they have called 'Lyapunov exponents'. Nonetheless, the dual transfer matrix product grows exponentially with system size (e.g., [46]), not solely because of uncertainty growth, so aren't related to classical chaotic dynamics or global Lyapunov exponents even if there may be a connection with the thermodynamic limit.

Recall that the failure to find the features of classical chaos in quantum systems usually is diagnosed as due to the linear nature of Schrödinger's equation because nearby states always remain close. However, some alternative possibilities for chaotic behavior have been proposed for interacting quantum systems.

Fred Kronz [47, 48] argues that the separable/nonseparable Hamiltonian distinction is more appropriate than nonlinearity regarding signatures of classical chaos in QM. Open quantum systems provide many examples of nonseparable Hamiltonians in QM. One important example is the set of Hamiltonians describing interaction between quantum systems and measurement apparatus. The quantum system-measurement apparatus compound system can evolve from a tensor product state to a nonseparable entangled state represented by an irreducible superposition of tensor product states. A second important example would be quantum entanglement.

Since Schrödinger's equation is no longer valid for interacting quantum systems, master equations are used instead (e.g., [49]). The latter have nonseparable Hamiltonians because the time evolution of the components of open systems is nonunitary. Such quantum systems have no formal constraint against states diverging far from each other. Moreover, interacting quantum systems have continuous spectra, a characteristic of classical macroscopic systems.

While these distinctives of interacting quantum systems sound more promising for classical chaotic dynamics, only universal statistical characteristics of energy spectra and fluctuations similar to those found in isolated systems have been observed (e.g. [24, 50, 51]). Open quantum systems have an enormous number of degrees of freedom due to dissipation or noise introduced by an environment. One usually traces over or integrates out the vast majority of these degrees of freedom. Still, these math procedures don't adequately resolve the physical mismatches between potentially infinite degrees of freedom in interacting quantum systems versus the finite degrees of freedom possessed by classical systems.

The quantum chaos literature often uses a broader notion of chaos as behavior that 'cannot be described as a superposition of independent one-dimensional motions' [51], p 357. While this is a form of inseparability, the behavior in open quantum systems is similar to that observed in isolated systems: universal statistical patterns sharing some relationship with classical chaotic counterparts as found in isolated systems (section 8.3).

Some work in driven quantum systems seems to show a relationship between pitchfork bifurcations of some classical Hamiltonian dynamical systems and changes to ground state entanglement in their corresponding semiclassical analogs (e.g., [52–54]). However, quantum pitchfork bifurcations in open quantum systems don't lead to a sequence of period doublings as in classical chaotic dynamics that we know of. Even though the quantum behavior bears no relationship to classical chaotic dynamics, interestingly, interacting quantum systems can exhibit a phase transition at a critical point corresponding to parameter magnitudes where semi-classical counterpart systems exhibits a pitchfork bifurcation of its fixed point attractor. This quantum critical point is where peak entanglement with respect to parameters, such as the coupling strength between quantum spins, occurs.

The Bose–Hubbard model for N interacting bosons is a well-studied model interacting system (e.g., [55]). The master equation based on this Hamiltonian is deterministic. In Bose–Hubbard systems, bosons hop from site to site such as atoms hoping from one lattice point to another or photons tunneling from one optical cavity to another. Creation and annihilation operators describe the dynamics as the annihilation of a boson at a site and the creation of a boson at a neighboring site. Bosons hop from site to site another probabilistically independent of one another. Therefore, there is a finite probability of all, none, or any number in between of bosons at any given site at any time. This implies a finite probability bosons will interact with each other.

Often, a laser is used to drive the system, which can be manipulated to produce a context where bosons are more localized. Such manipulations can reduce the amount of hopping by lowering the probability of boson interactions. By using a mean-field approximation, the multi-site Hamiltonian can be reduced to an effective single-site Hamiltonian. Quantum dynamics in the Bose–Hubbard model can then be related to a semiclassical model via an approximation where the quantum master equation is replaced by a semiclassical version (e.g., [56]).

For the stationary state case without driving, a change from a distribution centered at one location to a bimodal distribution occurs as the interaction strength among bosons is increased. Such a transition resembles a pitchfork bifurcation, and both the master and mean-field equations exhibit this transition. Furthermore, the transition in the mean-field case can be related to the pitchfork bifurcation from one stable attractor to a stable period two attractor in the corresponding semiclassical model. Increasing the interaction strength further produces a bifurcation with three maxima for the boson occupation distribution so long as the tunneling amplitude is small. However, this quantum bifurcation has no semiclassical counterparts. Quantum bifurcations manifest as structural changes in the boson location occupations in the quantum ground state density matrix. Nonetheless, no instability in the quantum system's steady state is observed. Pitchfork bifurcations in classical chaos, such as the logistic map, signal when a stable fixed point becomes unstable, a striking contrast to the quantum and semiclassical behavior.

For periodic driving in the Bose–Hubbard model, classical counterparts manifests period doubling sequences leading to chaotic dynamics. Meanwhile, counterpart mean-field models can be driven through a series of period doublings producing a plot reminiscent of classical chaos. There are suggestive visual similarities in the plots of maxima in the diagonal matrix elements (compare figures 4(a) and (b) in [56]). In this work, Ivanchenko *et al* apply what is sometimes called the quantum trajectory method but actually is a Monte Carlo method for tracking quantum jumps [57] to resolve the details.[14] Calculations yield boson number expectation

---

[14] Despite the name 'quantum trajectory method,' there are no trajectories (sometimes a source of confusion in the literature).

values at the first site for different discrete moments in time. Sampling the data points at these discrete times yields a series of maxima and minima in the diagonal elements of the asymptotic density matrix resembling the structure of the mean-field approximation. Nevertheless, it's possible to drive a bifurcation in the master equation model that has no correspondence to either the mean-field or semiclassical counterparts.

As another example of an open quantum system, consider a charged particle in the unit square with periodic boundary conditions undergoing an occasional kick from an external electromagnetic field.[15] For some parameter magnitudes, the classical counterpart model exhibits stretching and folding dynamics, a necessary condition for chaos, has positive global Lyapunov exponents, and is algorithmically complex. For the quantum model, the electromagnetic kick transforms the quantum labels of wave functions initially close together to labels which don't necessarily ever come close again. Such behavior is reminiscent of the divergence of classical chaotic trajectories and changes in wave function labels lead to an absolutely continuous 'quasi-energy spectrum'.[16] Particle position expectation values become unpredictable with respect to the initial wave function labels after long times and the sequence of quantum label shifts are algorithmically complex. One can even define a 'distance' between the labels that increases exponentially with time.

This is the closest analogy in quantum chaos to classical chaos. However, the quantum chaos conjecture (section 8.3.2) is inapplicable to this system due to the continuous spectrum of the quasi-energy. Moreover, exponential divergence by itself is neither necessary nor sufficient to characterize a system as chaotic. While the dynamics of quantum labels for a kicked charged particle is irregular and may be algorithmically complex, actual temporal evolution of the wave functions is algorithmically compressible. There is no irregularity in the evolution of wave functions themselves for this system.

While the very interesting behavior explored in quantum chaos may lack the characteristics of classical chaotic dynamics, the former has actual-world implications. For example, a challenge in quantum computing is preventing information loss in quantum states due to decoherence such as entanglement with the environment. Protocols based on quantum chaos can counteract decoherence effects by spreading information among quantum states while preserving information from the initial state (e.g., [59]).

8.5 Quantum chaos and confusion

Fascinating as the work on interaction systems, such as Bose–Hubbard systems, is, it also offers ample opportunities for confusion in the literature illustrated by the following example. Consider a ring of three coupled cavities and a chiral drive for photons in the cavities [60]. The Bose–Hubbard Hamiltonian for photons with a driving term coherent with respect to both frequency and momentum can describe

---

[15] Mathematically, this model is a generalization of the quantized Arnold cat map [58].
[16] The quasi-energy is defined as the set of numbers representing the 'energy' in the evolution operator acting to transform the wave function labels.

this system. In a classical model developed through a mean-field approximation with weak to no photon–photon interaction, cavities become classical coupled nonlinear oscillators whose dynamics exhibit a positive global Lyapunov exponent and a strange attractor can appear due to dissipation in the system.

However, when photon interactions are too strong, no classical model can be approximated nor can the boson statistics of the photons be modeled by classical means. This fits with our intuitions about differences between classical and quantum systems. Photon occupation is tracked by a reduced density matrix for the coupled cavities. A Monte Carlo simulation can model the evolution of the boson system from its initial vacuum state. The simulations produce dense occupation of an attractor in state space. Furthermore, the OTOCs exhibits sensitive dependence on the strength of perturbations from the chiral drive as well as growing exponentially. The authors use the term 'classical chaotic basin' of attraction without any discussion of the differences between the classical and quantum state spaces (the latter cannot support strange attractors), the lack of trajectories in the latter, or even the meaning of correlation functions in the case of classical trajectories versus discrete hopping of photons from one cavity to another in the quantum case.

This is an example of how confusing classical and quantum concepts and observables can lead to inappropriate characterizations of quantum behavior. What the authors can only mean is that some of the characteristics of quantum chaos show up in their quantum system under chiral driving while the analog classical system, living in a different state space with a different algebra of observables, exhibits marks of chaotic dynamics. Nevertheless, it's easy to read [60] as describing classical chaos showing up in a QM system because of the lack of clarity.

Although many more examples of confusing classical and quantum chaos concepts and properties could be given, this case provides a warning to treat quantum and classical systems and their distinctions with clarity and care.

References

[1] Berry M V 1987 Quantum chaology *Proc. R. Soc.* A **413** 183–98
[2] Berry M V 1989 Quantum chaology, not quantum chaos *Phys. Scr.* **40** 335–6
[3] Van Vleck J H 1928 Correspondence principle in the statistical interpretation of quantum mechanics *Proc. Natl Acad. Sci. USA* **14** 178–88
[4] Gutzwiller M C 1990 *Chaos in Classical and Quantum Mechanics* (New York: Springer) Interdisciplinary Applied Mathematics https://doi.org/10.1007/978-1-4612-0983-6
[5] Batterman R W A 2021 *A Middle Way: A Non-Fundamental Approach to Many-Body Physics* (New York: Oxford University Press) https://doi.org/10.1093/oso/9780197568613.001.0001
[6] Bishop R C 2024 *The Physics of Emergence* 2nd ed (Bristol: IOP Publishing) https://doi.org/10.1088/978-0-7503-6367-9
[7] McComb W D 2004 *Renormalization Methods: A Guide for Beginners* (Oxford: Clarendon) https://doi.org/10.1093/oso/9780198506942.001.0001
[8] Berry M V *et al* 1979 Quantum maps *Ann. Phys.* **122** 26–63
[9] Tomsovic S and Heller E J 1993 Long-time semiclassical dynamics of chaos: the stadium billiard *Phys. Rev.* E **47** 282–300

[10] Ge Z *et al* 2024 Direct visualization of relativistic quantum scars in graphene quantum dots *Nature* **635** 841–6

[11] Casati G and Prosen T 2005 Quantum chaos and the double-slit experiment *Phys. Rev. A* **72** 032111

[12] Berry M V 1977 Regular and irregular semiclassical wave functions *J. Phys. A* **10** 2083–198

[13] Bunimovich L A 2019 Physical versus mathematical billiards: from regular dynamics to chaos and back *Chaos* **29** 091105

[14] García-Mata *et al* 2023 Out-of-time-order correlations and quantum chaos *Scholarpedia* **18** 55237

[15] Rozenbaum E B *et al* 2020 Early-time exponential instabilities in nonchaotic quantum systems *Phys. Rev. Lett.* **125** 014101

[16] Larkin A and Ovchinnikov Y N 1969 Quasiclassical method in the theory of super-conductivity *Sov. Phys. JETP* **28** 1200

[17] Rozenbaum E B *et al* 2017 Lyapunov exponent and out-of-time-ordered correlator's growth rate in a chaotic system *Phys. Rev. Lett.* **118** 086801

[18] Hummel Q *et al* 2019 Reversible quantum information spreading in many-body systems near criticality *Phys. Rev. Lett.* **123** 160401

[19] Pilatowsky-Cameo S *et al* 2020 Positive quantum Lyapunov exponents in experimental systems with a regular classical limit *Phys. Rev. E.* **101** 010202

[20] Shepelyanskii D L 1981 Dynamical stochasticity in nonlinear quantum systems *Theor. Math. Phys.* **49** 925–8

[21] Gutzwiller M C 1971 Periodic orbits and classical quantization conditions *J. Math. Phys.* **91** 343–58

[22] Berry M V 2001 Chaos and the semiclassical limit of quantum mechanics (is the Moon there when somebody looks?) *Quantum Mechanics: Scientific Perspectives on Divine Action* ed R J Russell, P Clayton, K Wegter-McNelly and J Polkinghorne (Castel Gandolfo: CTNS Publications Vatican Observatory) 41–54

[23] Wigner E P 1951 On the statistical distribution of the widths and spacings of nuclear resonance levels *Math. Proc. Camb. Phil. Soc.* **47** 790–8

[24] Guhr T *et al* 1998 Random-matrix theories in quantum physics: common concepts *Phys. Rep.* **299** 189–425

[25] Wigner E P 1955 Characteristic vectors of bordered matrices with infinite dimensions *Ann. Math.* **62** 548–64

[26] Wigner E P 1967 Random matrices in physics *SAIM Rev.* **9** 1–23

[27] Casati G *et al* 1980 On the connection between quantization of nonintegrable systems and statistical theory of spectra *Lett. Nuovo Cimento* **28** 279–82

[28] Bohigas O *et al* 1984 Characterization of chaotic quantum spectra and universality of level fluctuation laws *Phys. Rev. Lett.* **52** 1–4

[29] Fyodorov Y 2011 Random matrix theory *Scholarpedia* **6** 9886

[30] Percival I C 1973 Regular and irregular spectra *J. Phys. B* **6** L229–32

[31] Berry M V and Tabor M 1977 Level clustering in the regular spectrum *Proc. R. Soc. A* **356** 375–94

[32] Sieber M and Steiner F 1990 Quantum chaos in the hyperbola billiard *Phys. Lett. A* **148** 415–20

[33] Andreev A V *et al* 1996 Semiclassical field theory approach to quantum chaos *Nucl. Phys. B* **482** 536–66

[34] Müller S *et al* 2009 Periodic-orbit theory of universal level correlations in quantum chaos *New J. Phys.* **11** 103025

[35] Altland A *et al* 2015 A review of sigma models for quantum chaotic dynamics *Rep. Prog. Phys.* **78** 086001

[36] Ullmo D 2016 Bohigas-Giannoni-Schmit conjecture *Scholarpedia* **11** 31721

[37] Rigol M *et al* 2008 Thermalization and its mechanism for generic isolated quantum systems *Nature* **452** 854–8

[38] Hallam A *et al* 2019 The Lyapunov spectra of quantum thermalisation *Nat. Commun.* **10** 2708

[39] Zhilinskii B I 2001 Symmetry, invariants, and topology II: symmetry, invariants, and topology in molecular models *Phys. Rep.* **341** 85–171

[40] Zhilinskii B I 2009 Quantum bifurcations *Encyclopedia of Complexity and Systems Science* ed R Meyers (New York: Springer) 7135–54

[41] Pierre G *et al* 1989 Organization of quantum bifurcations: crossover of rovibrational bands in spherical top molecules *Europhys. Lett.* **10** 409–14

[42] Bayfield J E and Koch P M 1974 Multiphoton ionization of highly excited hydrogen atoms *Phys. Rev. Lett.* **33** 258–61

[43] Casati G *et al* 1979 Stochastic behavior of a quantum pendulum under a periodic perturbation *Stochastic Behavior in Classical and Quantum Hamiltonian Systems, vol 93 of Lecture Notes in Physics* (Berlin: Springer) 334–52

[44] Fishman S *et al* 1982 Chaos, quantum recurrences, and anderson localization *Phys. Rev. Lett.* **49** 509–12

[45] Chirikov B V *et al* 1988 Quantum chaos: localization versus ergodicity *Physica* D **33** 77–88

[46] Chan A *et al* 2021 Spectral Lyapunov exponents in chaotic and localized many-body quantum systems *Phys. Rev. Res.* **3** 023118

[47] Kronz F M 1998 Nonseparability and quantum chaos *Phil. Sci.* **65** 50–75

[48] Kronz F M and Tiehen J T 2002 Emergence and quantum mechanics *Phil. Sci.* **69** 324–47

[49] Davies E B 1976 *Quantum Theory of Open Systems* (MA: Academic)

[50] Filikhin I *et al* 2011 Disappearance of quantum chaos in coupled chaotic quantum dots *Phys. Lett.* A **375** 620–3

[51] Ponomarenko L A *et al* 2008 Chaotic Dirac billiard in graphene quantum dots *Science* **320** 356–8

[52] Hines A *et al* 2005 Quantum entanglement and fixed-point bifurcations *Phys. Rev.* A **71** 042303

[53] Nemes M C *et al* 2006 Quantum entanglement and fixed point Hopf bifurcation *Phys. Lett.* A **354** 60–6

[54] Schneider S and Milburn G J 2002 Entanglement in the steady state of a collective-angular-momentum (Dicke) model *Phys. Rev.* A **65** 042107

[55] Le Boité A *et al* 2014 Bose-Hubbard model: relation between driven-dissipative steady states and equilibrium quantum phases *Phys. Rev.* A **90** 063821

[56] Ivanchenko M *et al* 2017 Quantum bifurcation diagrams *Ann. Phys. Lpz.* **529** 1600402

[57] Plenio M B and Knight P 1998 The quantum-jump approach to dissipative dynamics in quantum optics *Rev. Mod. Phys.* **70** 101–44

[58] Arnold V I and Avez A 1968 *Ergodic Problems of Classical Mechanics* (Reading, MA: W A Benjamin)

[59] Harris J *et al* 2022 Benchmarking information scrambling *Phys. Rev. Lett.* **129** 050602

[60] Dahan D *et al* 2022 Classical and quantum chaos in chirally-driven, dissipative Bose–Hubbard systems *npj Quantum Inf.* **8** 14

IOP Publishing

An Introduction to Chaotic Dynamics
Classical and quantum
Robert C Bishop

Chapter 9

Chaos and the classical–quantum relationship[1]

The distinctive differences between classical and quantum chaos already demonstrate important discontinuities between macroscopic and microscopic domains. Although many textbook discussions treat the relationship between these two domains as reductive—for instance, quantum mechanics (QM) supposedly reduces classical mechanics in some limit—our tour of quantum chaos indicates the relationship between microscopic and macroscopic domains is more subtle than this simple picture. This brings the focus to the venerable correspondence principle and its meaning.

9.1 Chaos and failure of the correspondence principle

As discussed in the previous chapter, important features of classical chaos, such as sensitive dependence on initial conditions (SDIC), exponential growth in uncertainty, the period doubling route to chaos, aperiodic trajectories, and apparent randomness, among others, are absent from quantum systems. There are interesting quantum chaos signatures in quantum systems when the corresponding classical system exhibits chaotic dynamics (e.g., [2, 3]). While the quantum signatures don't resemble those of classical chaos, the statistics of random matrices, repulsion among energy levels, specific forms of energy level dynamics, and growth in quantities such as out-of-time-ordered correlators (OTOCs) or entanglement in ground states are of interest in their own right. However, many of these show up only in semiclassical chaotic systems rather than strictly quantum systems.

The strong distinction between quantum and classical chaos is not unlike the distinct kinds of properties QM has with respect to classical physics. These distinctions have led to arguments that the correspondence principle between quantum and classical physics fails or that the former may be incomplete (e.g., [4]). As Max Jammer has pointed out, 'there was rarely in the history of physics a comprehensive theory which owed so much to one principle as quantum mechanics owed to Bohr's correspondence principle' [5], p 118. Hence, any challenge classical

---

[1] The material discussed in sections 9.1–9.3 has been discussed from a philosophical perspective in [1].

doi:10.1088/978-0-7503-6453-9ch9

chaos raises for the correspondence principle is important to understand as it's relevant to the relationship between QM and macroscopic physics.

Roughly, the correspondence principle is understood in two ways in the literature. The broader understanding is that scaling up a quantum system's size should lead to behavior that becomes classical at macroscopic scales. The narrower understanding is that in the limit of large quantum numbers, the behavior of a quantum system should reproduce the behavior of macroscopic classical systems.

Whether broadly or narrowly understood, the correspondence principle commonly is described in textbooks and articles as involving Planck's constant going to zero, perhaps in conjunction with relevant quantum numbers becoming very large (i.e., $n \to \infty$), as in [6]. But such approaches are problematic given that h is an unchanging constant of nature (e.g., chemistry would fail if its magnitude changed slightly). More rigorously, one should focus on relevant limits of ratios of the classical to quantum actions, for instance. However, all these limits are singular [7–9]. Quantum systems don't become increasingly similar to macroscopic systems as quantum numbers get large as illustrated by semiclassical system behavior (section 8.2). Instead, the transitions between quantum and classical properties and behaviors are abrupt.

Joseph Ford and Giorgio Mantica argue that 'any two valid physical theories which have an overlap in their domains of validity must, to relevant accuracy, yield the same predictions for physical observations'.[2] Understanding the correspondence principle this way implies 'quantum mechanics must, in general, agree with the predictions of Newtonian mechanics when the systems under study are macroscopic' [4], p 1087. For Ford and Mantica, 'The very essence of correspondence lies in the notion that quantum mechanics can describe events in the macroscopic world without any limit taking. Were this not the case, then there would be no overlap in the quantal and classical regions of validity' [4], p 1088. Sir Michael Berry puts it this way: 'all systems,' even the Moon, 'obey the laws of quantum mechanics' [10], p 42.[3] Therefore, 'if there is chaos (however defined) in the macroscopic world, quantum mechanics must also exhibit precisely the same chaos, else quantum mechanics is not as general a theory as popularly supposed' [4], p 1088.

Gordon Belot and John Earman [11] argue that a weak enough correspondence principle should only apply to physically realistic classical models. Moreover, under such a weakened principle only predictions of the quantized quantum counterparts applied to macroscopic situations should agree within experimental error with classical models. Suitably weakened, the correspondence principle requires QM 'reproduce to within experimental error the verifiable predictive successes of [classical mechanics] in macroscopic experiments on actual physical systems.' Hence, QM should not be required to reproduce results of all possible classical models. Failure of this weakened correspondence principle would 'be a direct challenge to either the completeness or the truth of QM' (p 163).

---

[2] The phrase 'an overlap' has an ambiguity as to whether the overlap is total or not. Ford and Mantica assume total overlap, but if two physical theories have partial, nontrivial overlap, there is no expectation that they should make the same physical predictions. More on this below.

[3] Notice that the governance myth (section 1.9.1) is presupposed in Berry's claim.

Belot and Earman offer two cases to resolve the problem. The first we've already met: in some quantum systems, wave packets mirror classical trajectories for very short times while exhibiting exponential divergence from each other. As discussed in chapters 5 and 8, however, mere exponential divergence by itself isn't a reliable indicator of chaotic behavior. Furthermore, many if not most of these time scales are too short to be relevant for macroscopic systems.

Their second case is that of long times (i.e., in the $t \to \infty$ limit), where, in an appropriate limit, strong mixing emerges. However, strong mixing also is neither necessary nor sufficient by itself to distinguish chaotic behavior (section 5.3 and appendix A).

Given that classical chaotic behavior isn't recovered in QM, the dilemma Ford and others describe looks serious: either the correspondence principle is false or QM is incomplete. Ford and Mantica, Berry, among others, reject the first horn of the dilemma. Therefore, the problem must lie with QM: its lack of classical chaos reveals some incompleteness in the theory. But are these the only two viable options for understanding the relationship between quantum and classical chaos and the correspondence principle?

9.2 Chaos and the failure of the false dilemma

How Ford and Mantica, Berry, among others, frame the problem presupposes a common misconception of the relationship between the quantum and classical domains leading to a false forced choice: QM should fully reproduce classical phenomena. Hence, the worry that QM is incomplete.

A more realistic approach to the quantum–classical relationship can dissolve the false forced choice. The first point is recognizing that this relationship involves a series of limits of the ratios of quantum observables involving Planck's constant and classical observables, such as relevant classical and quantum actions. As well, there are limits involving the separation of nuclear and electronic frames of motion (in the case of chemistry) among others. These limits all involve singular asymptotic series.[4] This fact undermines the reductionist assumption underlying the false forced choice because there is a qualitative change in the character of the states and observables going from quantum to classical domains. Classical states, observables, and their associated state spaces aren't functions of nor straightforwardly related to the states, observables, and state spaces intrinsic to QM (e.g., [14, 15]).

Second, different macroscopic worlds result from taking these various limits in different orders. These limits correspond to different physical transitions. Changing the order of the limits—changing the order of physical transitions—yields physically inequivalent macroscopic domains, only one of which corresponds to the classical physics of our experience.

This circumstance means one has to rethink the 'approximately classical' or 'quasi-classical' trajectories for quantum systems derived from semiclassical

---

[4] Hence, the relationship between quantum and classical phenomena isn't one involving anything like bridge laws relating the two domains as typical reduction accounts envision (e.g., [12, 13]).

considerations (e.g., [10, 16]). For instance, such quasi-classical behavior is exhibited only for limited times (except for overly idealized models) and under very special initial conditions for ground states only [17], p 166. Classical behavior is never exhibited by excited energy eigenstates. Appeal to Ehrenfest's theorem only guarantees that for such short-lived dynamics the usual physics practice of averaging the magnitudes of the quantum observables tends to wash out the errors or differences between the classical and quantum calculations for contextually relevant situations and times. The theorem shouldn't be interpreted as providing either necessary or sufficient for macroscopic behavior. Consider the application of Ehrenfest's theorem to a quantum harmonic oscillator. The resulting average quantities for the position and momentum may track with classical macroscopic quantities for some brief time, but the quantum oscillator's discrete states yield thermodynamic properties at variance with the corresponding classical oscillator.

Third, while some argue the emergence of our classical world is simply a matter of environmental decoherence (e.g., [10, 18]), this isn't the case. There is no context-free limit of infinitely many degrees of freedom. Such limits always have uncountably infinitely many physically inequivalent representations and context determines which potential representations are actualized [14, 15]. Moreover, in contrast to [19], an improper mixture of quantum states doesn't licence an inference that these states can be treated as satisfying a classical probability distribution. Such mixed states aren't effectively classical, meaning any system described by a nonpure density operator isn't even approximately equivalent to a classical 'mixture' with an exact state an observer is ignorant about. One can interpret impure quantum states as classical mixtures if and only if the components of the former are described by disjoint states. For a classical mixture of two pure states, the pure states are disjoint if and only if there exists a classical observable such that the expectation values with respect to these states are different. This is a key distinction between classical and quantum states [20].

What about Bohmian mechanics as an approach to QM's relationship to chaos? David Bohm [21] proposed a separation of Schrödinger's equation into a set of coupled equations describing how particle trajectories behave guided by a quantum potential. To ensure the gradient of the phase of the wave function remains single-valued, he restricted solutions to a subset satisfying a continuity equation.[5]

The issue with the relationship between the quantum and classical domains remains, however, and one shouldn't expect everything in classical mechanics to be captured by a Bohmian approach (e.g. [22]). For Bohm's approach, the 'classical limit' is when the quantum potential goes to zero, making negligible contributions to classical particle behavior. The quantum potential depends on a system's wave function, therefore not all quantum systems have a 'classical limit' in this sense corresponding to a macroscopic system.

---

[5] Otherwise, the set of solutions yields statistical predictions at odds with QM.

Furthermore, the emergence of a classical realm resulting as the quantum potential goes to zero is misleading. The ontology of Bohm's theory already represents an algebra of observables for particles that are classical (e.g., position and momentum for particle trajectories). The quantum potential does all the 'quantum work,' with the wave function serving to guide the classical trajectories. In other words, Bohmian mechanics already involves mixed quantum/classical algebras of observables (i.e., having classical observables as their center among quantum observables). While, there are examples of chaotic trajectories in Bohmian mechanics, the corresponding classical systems in the limit when the quantum potential goes to zero exhibit regular trajectory behavior (e.g., [23]).

There has been work searching for aperiodic trajectories that diverge from each other exponentially quickly in Bohmian mechanics (e.g., [24–26]). Given the quantum/classical nature of Bohm's theory and the presence of classical trajectories, it's unsurprising that SDIC might appear among the classical observables describing the trajectories for interacting systems.

Nevertheless, Bohmian particle trajectories are undetectable at quantum length scales, raising questions about how these relate to putative marks of classical chaos. Moreover, Bohmian quantum systems generally lack marks of quantum chaos (an effect of the mixed quantum/classical algebra of observables). Perhaps most provocative, Bohmian mechanics is an explicitly deterministic and nonlinear theory, but the chaotic dynamics it exhibits isn't produced by stretching and folding mechanisms in contrast to classical dynamical systems. Instead, trajectories coming close to nodal points, where the equations of motion become singular when the wave function equals zero, is what gives rise to SDIC [23]. Hence, Bohm's theory has an important contrast with the chaotic dynamics of classical systems raising questions about classical physical systems as limiting cases of Bohmian systems as the quantum potential goes to zero.

For the first horn of the dilemma, then, there is nothing in the quantum domain by itself fixing the character of the classical domain, though QM provides some necessary conditions for the latter. Classical chaos, along with many other classical features, is emergent in a more complex sense than assumed by Ford and others [14, 15]. This implies dropping the implicit assumption of reductionism (along with the governance myth) in these discussions of the relationship between quantum and classical chaos. Abandoning this assumption means the disparity between quantum and classical chaos no longer calls an appropriately formulated, nuanced correspondence principle into question (see below). This resolves the first horn of the dilemma.

For the second horn of the dilemma, it's worth recalling that the state spaces of dynamical systems of classical physics and Hilbert spaces of QM are qualitatively distinct: different observables, different dynamics, different state spaces. This isn't a sign that QM is incomplete. Rather, it's a sign that the relationship between quantum and classical domains is different than Ford and others assume (section 9.4 and [15]).

Therefore, we don't have to choose between 'either the completeness or the truth of QM' [11], p 163.

9.3 Subtlety of the quantum–classical relationship

Another proposal for resolving the quantum–classical issue with the correspondence principle is to switch to the Koopman approach [11, 27, 28]. This amounts to everything quantum and classical being defined in a suitable Hilbert space. However, this trades particle trajectories for probability distributions and their dynamics. All the properties and tools for characterizing chaos become irrelevant; all that remains are the dynamics of energy distributions, random matrices, diffusion processes and so forth. Classical chaotic dynamics is lost when only the properties and tools of quantum chaos are left. As we've seen, these are insufficient to describe classical chaotic dynamics. Furthermore, the Koopman approach misses out distinctions between classical and quantum states, observables, and state spaces. The quantum–classical relationship is too subtle to be resolved by such a mathematical move.

The following two examples illustrate some of this subtlety and confusions that can arise when it's ignored.

9.3.1 Born–Oppenheimer systems and chaos

The Born–Oppenheimer procedure [29], among other adiabatic procedures, implements a stability condition corresponding to the physical significance of nuclear mass, namely that nuclear motion is much slower than that of electrons due to mass differences [15]. This nuclear mass stability condition involves a singular asymptotic series in the ratio of electron to nuclear mass. The nucleus is treated as if it's almost stationary with respect to electron motions resulting in breaking entanglements between electron and nuclear frames, distinguishing electrons from nucleons, and breaking permutation, rotational, and translational symmetries, all physical effects. Indeed, this 'clamped nucleus' assumption at the heart of Born–Oppenheimer and other adiabatic procedures is the only known approach yielding self-adjoint Hamiltonians for quantum chemistry [30]. Such systems are characterized by mixed quantum/classical algebras of observables characterizing the motion of nucleons and electrons.

The resulting quantum/classical systems are nonlinear, holding out a possibility for chaos. Poincaré surface of section plots can now be defined and used to study a one-dimensional molecule's position and momentum phase space, say. Blümel and Esser [31] applied a harmonic potential with appropriate parameter magnitudes to search for evidence of chaotic dynamics for molecular motion. But they don't offer any of the usual classical marks for chaotic dynamics, nor even distinguish classical from quantum dynamics. Instead, they turn for verification to the quantum observables. The amplitudes for different initial conditions are plotted on the **Bloch sphere**, a representation of quantum states and their dynamics (imagine arrows originating in the middle of a sphere with tips touching the surface). These amplitudes diverge from each other with exponential separation characterized by a positive 'Lyapunov exponent' on the Bloch sphere. Blümel and Esser never explain why exponential divergence in state amplitude is an appropriate analog for Lyapunov exponents nor how it relates to macroscopic contexts (given that quantum observables don't permit appropriate notions of trajectory associated with such exponents). The appellation is given solely because the amplitude divergence is exponential. They are dealing with an emergent quantum/classical algebra of observables, but only clearly demonstrate a characteristic from the quantum chaos zoo.

The differences between classical and quantum systems is apparent as classical observables, treated as a dynamical subsystem, exhibit marks of classical chaos while quantum observables don't.

9.3.2 Quantum tunneling and chaos

Although tunneling is a quintessential quantum phenomenon, there is a surprising link between chaotic dynamics in classical models and tunneling in quantum counterparts. A textbook case is a classical particle in a double-well potential. For no driving force, a low energy particle is confined to one well of the potential. For high enough energies the particle may cross over the potential barrier into the other well. There is a natural separation of the low-energy from high-energy orbits in the position-momentum phase space. When an oscillating driving force is applied, a chaotic layer appears in phase space, the layer's size depending on the amplitude and frequency of the driving force. This layer indicates which orbits of various energies cross from one well to the other [32].

As is well known, in the quantum textbook case, while the energy barrier is classically impenetrable, quantum tunneling leads to a finite probability for particles to pass from one well to another even at the lowest energies. Applying a periodic driving force to the semiclassical counterpart leads to a surprising tunneling rate significantly greater than the textbook case in those regions of the phase space of the semiclassical model corresponding to the aperiodicity of the chaotic layer in the classical counterpart [33]. Chaotic dynamics in the classical phase space determines the tunneling rate in the corresponding semiclassical model: particles starting in regions of the semiclassical phase space corresponding to regular orbits in the classical model exhibit normal tunneling rates.

However, quantum orbits in regions corresponding to classically chaotic orbits exhibit significantly higher tunneling rates. If quantum particles begin in semiclassical phase space regions corresponding to classically regular orbits, they can tunnel to other regions corresponding to classically regular orbits. Quantum particles avoid any regions associated with classical chaotic dynamics, maintaining their coherence oscillating back and forth between regions corresponding to classically regular orbits. These particles never spread into regions corresponding to classically chaotic orbits even if there is a large region of chaotic behavior in the corresponding classical phase space.

More complex semiclassical models involving multiple degrees of freedom such as two coupled quartic oscillators exhibit similar behavior (e.g., [34]).[6] For the classical case, zero coupling between the two oscillators yields regular oscillatory motion. Increasing the coupling strength causes the system to enter a quasi-integrable

---

[6] The double-well potential is a one-degree-of-freedom system.

regime. Chaotic regions in state space begin to appear. As the coupling strength increases further, the system reaches a regime where the state space exhibits widespread aperiodic orbits and SDIC.

For the semiclassical quantum analog, tunneling between the two quantized oscillators is affected in regions of the state space corresponding to chaotic dynamics in the classical model state space as the coupling increases and as Planck's constant 'increases' from zero. Erratic energy splittings for the discrete energy levels of the oscillators appear, where the statistics are described by random matrices. Again, the chaotic dynamics of the classical model influences the tunneling in the counterpart semiclassical quantum model.

9.4 Contextual emergence and the validity of quantum mechanics

As Gutzwiller observed in 1990, 'the breakdown of classical mechanics is probably more subtle and remote than most of the theoretical discussions so far' [2], p 175. Shifting from a classical chaotic model to semiclassical model to a fully quantum model and back by substituting quantum operators for classical observables or letting the value of h go to zero paper over the significant differences among states, observables, and state spaces in these three kinds of models and their physical domains. Hence, these techniques obscure some of the subtlety Gutzwiller was pointing towards, differences between quantum and classical chaos evince.

However, there is no reason to suspect that the issue is one of inadequacy in QM highlighted by chaos if classical or macroscopic physics is **contextually emergent** from the quantum domain [14, 15]. Unless, that is, one is wedded to reductionism, an implicit assumption in much of the literature on quantum chaos. Contextual emergence is defined as

Definition 9.1. Contextual emergence: properties and behaviors at a smaller scale or underlying domain (including its laws) offer *some necessary but no sufficient conditions* for properties and behaviors at a larger scale. Larger scales or domains provide the needed extra conditions.

This contrasts with reduction:

Definition 9.2. Reduction: properties and behaviors at a smaller scale or underlying domain (including its laws) offer *both necessary and sufficient conditions* for properties and behaviors at a larger scale.

For reduction, states, observables, and laws of the smaller scale or underlying domain are either identical with or imply all states, observables, and laws at a larger-scale or target domain. A standard textbook example of reduction is special relativity reducing to Newtonian mechanics in the limit $v^2/c^2 \rightarrow 0$.[7]

---

[7] The spacetime of special relativity doesn't reduce smoothly to the spacetime of Newtonian mechanics in this limit. This textbook example ignores some subtleties with reductionism.

In contrast, contextual emergence characterizes cases where states, observables, and laws of the smaller scale or underlying domain contribute some necessary but no sufficient conditions for states, observables, and laws or even systems in the larger scale target domain. For example, the domain of elementary particles and forces contributes some of the necessary conditions for wine flowing from a bottle into a glass: no elementary particles, no wine, bottle, or glass. Nonetheless, elementary particles and their forces don't guarantee wine exists or will be flowing out of a bottle into a glass. A complete set of necessary and sufficient conditions for wine and wine flow is contingent rather than necessary, involving more conditions than are found in the domain of elementary particles and forces.[8]

Stability conditions are crucial to defining the contexts in which elementary particles and forces come to expression, ensuring the existence and persistence of relevant states and observables as well as systems relevant states and observables characterize. Moreover, stability conditions make some behaviors available that previously were forbidden and vice versa. An example would be the large-scale structures in Rayleigh–Bénard convection providing the context shaping the behavior of the fluid parcels composing convective cells (see [15], section 4.1). Another example would be the 'clamped' nuclei of adiabatic procedures, such as Born–Oppenheimer, providing stability conditions for the emergence of molecular structure and nuclear and electronic equivalence classes (see [15], section 4.3). Such adiabatic procedures mark the context in which relevant states and observables making molecular shape an actual rather than possible phenomenon.[9]

In quantum chaos studies, the semiclassical limit $h \to 0$ points to a stability condition where relevant classical states and observables emerge with dynamics distinct from that of QM. And as with the transition from QM to chemistry, the transition from QM to semiclassical mechanics to classical mechanics involves singularities masked over in the substitution of quantum observables by classical ones or 'letting Planck's constant decrease to zero.' These heuristic techniques signal a change in states and observables relevant for macroscopic systems and processes versus those relevant for microscopic systems and those transitions are always singular.[10]

---

[8] For more discussion of reduction, contextual emergence, case studies (such as the relationship between chemistry and QM or the emergence of temperature), and objections, see [14, 15].

[9] An interesting, relevant example involving dimensionality is the Poincaré–Bendixson theorem [35], p 389–403, which states that chaos is only possible for continuous systems in three or more spatial dimensions. Because of uniqueness properties, trajectories in one and two dimensions cannot cross which restricts the possible end states of continuous dynamical systems implying only fixed point, cycles, and spirals are possible —no chaos. Spatial dimension is a stability condition for the possibility of chaotic dynamics in continuous systems. In contrast, discrete systems, such as the logistic map, have no such dimensional stability condition. There are no continuous trajectories, only trajectories that jump from point to point. In this sense, the property of discreteness versus continuity is a stability condition contributing to the possibility for chaos in dynamical systems. Both spatial dimension and continuity/discreteness are examples of acausal stability conditions [14].

[10] Gutzwiller [2] refers to equations developed in this limit as 'approximations' to the corresponding quantum equations. But, as with the Born–Oppenheimer procedure, such language is misleading, masking the emergence of new states and observables not present in QM leading to mixed quantum–classical algebras of observables (e.g., [9, 15]).

I have never seen the intricacies of this transition noted in quantum chaos discussions; hence, perhaps it's not surprising that Ford and others take an oversimplified view of the correspondence principle and quantum–classical relationship. An implicit assumption of reductionism can be blinding to physical subtleties.

The consequence is missing out why the phenomenon of classical chaos and its tools of study are distinctive of and only definable for macroscopic systems with classical states and observables. This also explains some of the confusion in the quantum chaos literature, for instance, 'Lyapunov exponents' are discussed in quantum systems. A Lyapunov exponent is a classical concept and only definable for trajectories in state spaces relevant to classical dynamics. While the quantum states and observables haven't vanished—the electrons and nucleons still exist—the macroscopic systems and properties shape the motions of quantum constituents (e.g., as you scan across this page reading the text, the electrons in your eyes are carried along even while partially constituting your eyes).

Contextual emergence implies that the differences between classical and quantum chaos call neither the generality nor the validity of QM into question. Nor does the complex, subtle emergent relationship imply the quantum and classical domains are nonoverlapping or totally disjoint. Rather, the overlap between the quantum and classical is partial and nontrivial, in contrast to Ford's, Berry's and others's assumption of complete overlap. QM is universally applicable in providing some necessary conditions for the existence of the macroscopic world (no electrons, protons, and neutrons, no weather systems); hence, Ford and Mantica's, Berry's, Belot and Earman's worry that QM isn't a general theory misses the mark. The universality of QM contributing some of the necessary conditions for all macroscopic systems and behaviors in no way implies QM alone universally governs macroscopic behavior[11]; hence, those, such as Berry or David Wallace [18], who for very different reasons think all systems obey the laws of QM, also miss the mark.

The quantum realm contributes some of the necessary conditions for classical properties and behaviors, but no sufficient conditions. One indicator of this is that macroscopic physics depends on continuous trajectories of individual particles through spacetime, while QM depends on wave functions and probabilities. There are deep conceptual and qualitative differences between the classical and the quantum. Even if one pursues Bohm's version of QM, which has continuous particle trajectories in spacetime, there still are important conceptual as well as physical and metaphysical differences as noted above (e.g., the presence of an all-pervasive quantum potential in Bohm's theory).

The bottom line is that one shouldn't expect individual continuous classical trajectories to result from QM in contextually inappropriate limits nor that QM should exhibit the full range of classical behaviors, contrary to the dilemma posed by Ford and others. Rather, one should expect that quantum probabilities recover the classical probabilities for appropriate stability conditions, and that there should be some nontrivial and only partially overlapping relationships between quantum and

---

[11] Unless one is persuaded of the governance myth (section 1.9.1).

classical properties and behaviors. The interesting statistical regularities discovered in quantum chaos fit with this contextually emergent, nontrivial, partial overlapping relationship nicely. Hence, the distinct properties of quantum and classical chaos are features of a unified, coherent multi-scale understanding of the world [14, 15].[12]

References

[1] Bishop R C C 2024 *The Stanford Encyclopedia of Philosophy* (Winter 2024 Edition), E N Zalta & U Nodelman (eds.), https://plato.stanford.edu/archives/win2024/entries/chaos/

[2] Gutzwiller M C 1990 Chaos in Classical and Quantum Mechanics *Interdisciplinary Applied Mathematics* (New York: Springer) https://doi.org/10.1007/978-1-4612-0983-6

[3] Emary C and Brandes T 2003 Chaos and the quantum phase transition in the Dicke model *Phys. Rev.* E **67** 066203

[4] Ford J and Mantica G 1992 Does quantum mechanics obey the correspondence principle? Is it complete? *Am. J. Phys.* **60** 1086–97

[5] Jammer M 1966 *The Conceptual Development of Quantum Mechanics* (New York: McGraw-Hill)

[6] Van Vleck J H 1928 Correspondence principle in the statistical interpretation of quantum mechanics *Proc. Natl. Acad. Sci. USA* **14** 178–88

[7] Friedrichs K 1955 Asymptotic phenomena in mathematical physics *Bull. Am. Math. Soc.* **61** 485–504

[8] Dingle R 1973 *Asymptotic Expansions: Their Derivation and Interpretation* (New York: Academic)

[9] Primas H 1998 Emergence in exact natural sciences *Acta Poly. Scand.* **91** 83–98

[10] Berry M V 2001 Chaos and the semiclassical limit of quantum mechanics (is the Moon there when somebody looks?) *Quantum Mechanics: Scientific Perspectives on Divine Action* ed R J Russell, P Clayton, K Wegter-McNelly and J Polkinghorne (Castel Gandolfo: CTNS Publications Vatican Observatory) 41–54

[11] Belot G and Earman J 1997 Chaos out of order: quantum mechanics, the correspondence principle and chaos *Stud. Hist. Phil. Mod. Sci.* **28** 147–82

[12] Nagel E 1961 *The Structure of Science: Problems in the Logic of Scientific Explanation* (New York: Harcourt, Brace, and World)

[13] van Riel R and Van Gulick R 2024 *Scientific Reduction* (Stanford, CA: Stanford University) https://plato.stanford.edu/archives/spr2024/entries/scientific-reduction/

[14] Bishop R C et al 2022 *Emergence in Context: A Treatise in Twenty-First Century Natural Philosophy* (Oxford: Oxford University Press) https://doi.org/10.1093/oso/9780192849786.001.0001

[15] Bishop R C 2024 *The Physics of Emergence* 2nd ed (Bristol: IOP Publishing) https://doi.org/10.1088/978-0-7503-6367-9

[16] Berry M V 1987 Quantum chaology *Proc. R. Soc.* A **413** 183–98

---

[12] Contextual emergence and stability conditions help answer Sabine Hossenfelder's question 'that no one has an answer' to [36]: how can classical chaos emerge from QM? The short answer is that one has to do the hard, delicate work of following through every asymptotic limit involving the appropriate stability conditions to track the transitions from the quantum algebras of observables to the quantum/classical algebras of observables as discussed in [15], sections 4.2 and 4.3, for instance, to understand how the total set of necessary and sufficient conditions for classical chaotic behavior are related to QM. On the other hand, if one is committed to reductionism, as Hossenfelder is, then the question of relationship appears unanswerable.

[17] Pauli W 1933 *Die Allgemeinen Prinzipien der Wellenmechanik vol 24 of Handbuch der Physik* (Berlin: Springer) 83–272

[18] Wallace D 2012 *The Emergent Multiverse: Quantum Theory According to the Everett Interpretation* (Oxford: Oxford University Press) https://doi.org/10.1093/acprof:oso/9780199546961.001.0001

[19] Zurek W H 1991 Quantum measurements and the environment-induced transition from quantum to classical *Conceptual Problems of Quantum Gravity* ed A Ashtekar and J Stachel (Boston, MA: Birkhäuser) 43–62 pp

[20] Primas H 1983 *Chemistry, Quantum Mechanics and Reductionism: Perspectives in Theoretical Chemistry* Lecture Notes in Chemistry **vol 24** (Berlin: Springer) https://doi.org/10.1007/978-3-642-69365-6

[21] Bohm D 1952 A suggested interpretation of quantum theory in terms of 'hidden' variables, I and II *Phys. Rev.* **85** 166–93

[22] Holland P R 1993 *The Quantum Theory of Motion* (Cambridge: Cambridge Univ. Press) https://doi.org/10.1017/CBO9780511622687

[23] Contopoulos G and Tzemos A C 2020 Chaos in Bohmian quantum mechanics: a short review *Regul. Chaot. Dyn.* **25** 476–95

[24] Dürr D *et al* 1992 Quantum chaos, classical randomness, and Bohmian mechanics *J. Stat. Phys.* **68** 259–70

[25] Faisal F H M and Schwengelbeck U 1995 Unified theory of Lyapunov exponents and a positive example of deterministic quantum chaos *Phys. Lett.* A **207** 31–6

[26] Ivanov I A 2019 Quantum chaos in strong field ionization of hydrogen *J. Phys.* B **52** 225002

[27] Koopman B O 1931 Hamiltonian systems and transformations in Hilbert space *Proc. Natl Acad. Sci. USA* **17** 315–8

[28] Koopman J and von Neumann J 1932 Dynamical systems of continuous spectra *Proc. Natl Acad. Sci.* **16** 255–61

[29] Born M and Oppenheimer R 1927 Zur quantentheorie der molekeln *Ann. Phys.* **389** 457–84

[30] Sutcliffe B T and Guy Woolley R 2010 Atoms and molecules in classical chemistry and quantum mechanics *Handbook of the Philosophy of Science* **vol 6** ed R F Hendry *et al* (Amsterdam: Elsevier) 388–426 pp

[31] Blümel R and Esser B 1994 Quantum chaos in the Born-Oppenheimer approximation *Phys. Rev. Lett.* **72** 3658

[32] Reichl L E and Zheng W M 1984 Field-induced barrier penetration in the quartic potential *Phys. Rev.* A **29** 2186

[33] Lin W A and Ballentine L E 1990 Quantum tunneling and chaos in a driven anharmonic oscillator *Phys. Rev. Lett.* **65** 2927–30

[34] Tomsovic S and Heller E J 1993 Long-time semiclassical dynamics of chaos: the stadium billiard *Phys. Rev.* E **47** 282–300

[35] Coddington E A and Levinson N 1955 *Theory of Ordinary Differential equations* (New York: McGraw-Hill)

[36] Hossenfelder S Scientists uncover hidden pattern in quantum chaos, December 2024. https://www.youtube.com/watch?v=bA5l1OS58Jo.

Chapter 10

Broader implications

Chapter 6 traced some of the ways scientists have changed their approaches to modeling and forecasting in light of chaotic dynamics. While chaos has offered important lessons for forecasting weather and ecological systems, among others, it also affords opportunities for applications which are proving fruitful. More broadly, chaos raises questions about determinism, laws, causation, free will and divine action in the world when nonlinearities are present and active. Our intuitions about all of these topics have largely been shaped by thinking about linear systems and behaviors, so it's perhaps not surprising that chaos and nonlinear dynamics might reshape that thinking.

10.1 Chaos in the world

Aside from weather systems and forecasting (chapter 6), chaotic dynamics shows up in many areas in the actual world giving us insight and raising new questions.

Recall the logistic map (section 3.1), a simplified ecological model of one species population treated in isolation from any other organisms. In the actual world, species aren't isolated from one another. In laboratory environments, we can create contexts with one species and its food source, yet it's difficult to create conditions eliciting chaotic dynamics in the time series of population growth/decline resembling the behavior of the logistic map. Nevertheless, when adding populations of species to the mathematical model and laboratory context, richer chaotic behavior is observed due to interactions among species populations. Mathematical ecology models under these circumstances resemble the unpredictability observed in actual communities of organisms. Ensemble forecasting has proved promising for modeling more complex ecological systems (e.g., [1–3]). Such research has implications for management and conservation of ecosystems and threatened populations.

The stability of the Solar System has been of interest since Isaac Newton raised the question in the 17th century.[1] There have been attempts to use the Chirikov map (section 3.2) to model Solar System dynamics by assuming each time a planet comes to the point of closest approach to the Sun or to another planet, it receives a kick due to the force of gravity. However, gravity cannot be isolated to a single kick at a time.

Data on the trajectories of the planets and their masses are known to very good accuracy, hence we're able to run models of the Solar System using Newton's equations. This allows a determination of both past as well as future planetary orbits. Forecasts indicate that eight billion years in the future—roughly the expected age before the Sun expands to the point it swallows the inner planets—planetary orbits will look very similar to now.[2]

Nonetheless, when varying planetary masses or initial positions slightly within the uncertainties in our measurements (down to as small as a millimeter), it turns out that for time scales as short as five million years, Mercury can smash into Venus about one percent of the time in simulations. Hence, the Solar System is forecasted to be stable up to the death of the Sun with 99% probability, not unlike a modern weather forecast. This is a clear sign that Solar System models exhibit SDIC [5], and references therein. Recent work on studying chaotic dynamics provides a possible explanation for why the Solar System has persisted with remarkable stability and may do so for billions of years due to how the resonance harmonics interact with trajectories confining chaotic behavior along the path of planetary orbits leaving the shape of orbits largely stable [5].

From the larger scales to smaller scales, chaotic dynamics shows up in cells. Liquid-liquid phase separation is crucial to many biological processes. The resistance of water and oil to mixing is an example of liquid-liquid phase separation. Such separation is thermodynamically driven, meaning energy must be added to overcome the separation. This is why you must shake a bottle of vinaigrette salad dressing to mix ingredients before pouring it on your salad. Shaking provides the energy needed to break down liquid-liquid phase separation.

Modeling of such phase separation in cellular processes has been studied using polyethylene glycol (PEG) and dextran, a polysaccharide derived from glucose, adding in protein nanomotors to simulate active matter [6]. In the model system, PEG and dextran phase separate but the phase boundaries of the separation are about a thousand times weaker than oil and water. Hence, little added energy is needed to produce a mixture.[3] Adding protein nanomotors and microtubules to the system, the nanomotors cause the microtubules to flow past one another. This flow is chaotic under appropriate parameter magnitudes.

---

[1] He famously believed God had to intervene to restore order to planetary orbits from time to time: 'For while Comets move in very excentrick Orbs in all manner of Positions, blind Fate could never make all the Planets move one and the same way in Orbs concentrick, some inconsiderable Irregularities excepted, which may have risen from the mutual Actions of Comets and Planets upon one another, and which will be apt to increase, till this System wants a Reformation' [4], p 403.

[2] Because gravity is a long range force, over such a long time span the effects of all known comets, asteroids, and planetary moons, as well as passing stars, have had to be added into the models.

[3] This low energy barrier is a stability condition enabling a host of chemical processes in cells.

The nanomotors continuously stir the fluid system. If PEG and dextran are initially mixed, the nanomotors cause a phase separation leading PEG to form cohesive droplets within the dextran. The nanomotors and microtubules remain in the dextran, where microtubule flow moves the PEG droplets around. Under moderate nanomotor activity the PEG droplets begin coalescing. When nanomotor activity level is too high, larger PEG droplets break down as fast as the smaller droplets coalesce, leading to an equilibrium where droplets reach an upper limit in size. Microtubules start climbing walls against the force of gravity, where the chaotic dynamics of the microtubule flow deforms the PEG droplets-dextran interface and the liquid-solid interface between the liquids and the liquids and the wall. It is these deformations that are crucial to the motions of PEG droplets and the wall climbing processes observed in living cells. Chaos gives some insight into how active matter moves cellular constituents around.

10.2 Applications of chaos

Despite all the popularized talk about unpredictability, chaotic dynamics has many useful applications resulting from the forms of control that can be exerted over it. A key reason is that trajectories exhibiting chaos have well-defined, stable, and repeatable geometrical properties and frequency distributions. Systems can be designed to take advantage of such persistent features.

It has been recognized for several decades that various kinds of epileptic episodes occur when normally chaotic processes in brain centers become periodic [7]. Likewise, chaotic dynamics plays a role in the cardiovascular system [8], which may lead to early detection of heart disease [9]. Cancer biology is another area where chaotic dynamics in protein oscillations and other intracellular processes may play a role in unrestrained cell growth providing targets for possible interventions [10].

Physicists and engineers have studied networks of chaotic oscillators discovering that electronic oscillators can be manipulated simply by the voltage input into the system. Coupled oscillators interact in such a way that there is competition between attractors for system behavior versus the oscillators repulsing each other. The dynamics exhibited by chaotic networks of oscillators are sensitive to small changes which can be used as sensors for electronic signals or for distributed computation depending on the kind of voltage input [11].

The intricate structure of chaotic dynamics has led to chaos-based communications systems. For instance, a binary sequence of information can be coded by manipulating the time series produced by a chaotic system. This opens the door for communications performed with very low power apparatus since chaos can be produced in simple systems. Additionally, there has been recent work investigating the use of chaotic systems for secure communications (e.g., [12]).[4]

---

[4] See [13] for more examples.

10.3 Wider implications

The sensitive dependence and complex structure of chaotic dynamics has wider implications than just predictability limits and applications, however.

10.3.1 Chaos and determinism

As noted in section 1.3, chaos is a property of deterministic dynamical systems, but some have thought chaotic dynamics is indeterministic. Unique evolution is preserved in our mathematical models of chaos. There are limits to predictability, an epistemic property, due to chaos, but such limits provide no evidence for a breakdown of determinism, an ontic property [14, 15]. Any arguments linking unpredictability with failures of determinism confuse ontic and epistemic properties. Furthermore, although apparent indeterminism can be generated by using a coarse-grained state space to analyze chaotic behavior, this produces only apparent randomness (section 1.1) because the underlying equations remain fully deterministic.

Boris Chirikov, of Chirikov standard map fame, has claimed chaos 'destroyed the deterministic image of classical physics' [16], p 9. But is this meant in an epistemic or ontological sense? Epistemically, one might think this because the apparent randomness exhibited by chaotic dynamics doesn't look deterministic. The dynamics continues to be deterministic but appears to lack order.

Ontologically, distinctions between observables and parameters as well as between systems and boundaries are crucial for mathematical modeling. As illustrated throughout this book, the failure of linear superposition renders these distinctions problematic.[5] Changes to properties associated with these distinctions are relevant to contexts in which evolution equations can remain deterministic (e.g., [18–20]). Conditions enabling chaotic dynamics certainly are sufficient to disrupt contexts needed to guarantee unique evolution. However, if determinism is a necessary condition for chaos but is context dependent, then chaotic dynamics would also be context dependent. Once the stability conditions guaranteeing unique evolution are lost, chaotic dynamics would be lost too.

Questions raised by the faithful model framework (section 1.9) provide ontic possibilities for challenging the determinism of actual-world systems because of problems with the presumed connections between nonlinear dynamical systems and actual-world targets. Inferences from the deterministic character of our models to the deterministic character of actual-world systems may be weaker than assumed. The chaotic behavior of actual-world systems may be merely mimicked by our models rather than the latter telling us whether the former maintain unique evolution under all circumstances.

10.3.2 Chaos, laws, and causation

Beginning with the 17th century, laws of nature have been conceived of as the source of order in the Universe [21]. Chaos not only raises questions about determinism, but

---

[5] See also [17].

also about laws of nature. Additionally, there has been some debate about whether there are laws specific to chaotic dynamics (e.g., [22]).

There are two main traditions for accounts of laws of nature. One is the nomological or governance tradition. On these accounts, laws have metaphysical existence and enforce patterns and regularities on properties and processes in nature.[6] Prima face chaotic dynamics challenges the intuitions motivating the necessitarianism in these accounts. For example, chaotic behavior is sensitive to particular parameter magnitudes and exhibits SDIC. Nonetheless, the parameter magnitudes don't govern behavior any more than π governs the circumference of circles. Perhaps stretching and folding properties of chaos reflect underlying dynamical laws of systems. This would imply there are no genuine chaos laws over and above the various dynamical laws of classical physics subject to specific parameter magnitudes.

Second is the regularity tradition. On these accounts patterns and relations observed in nature exist with no additional metaphysical necessity or governance to enforce or sustain them. These accounts differ based on assessment of which regularities count as laws, such as best systems or supervenience approaches [23]. Chaotic dynamics exhibits some form of determination. For instance, the chaos of classical models plays some determining role in the behavior of the corresponding quantum models, such as modifying tunneling rates (section 9.3.2). Yet, these determination or shaping relations run opposite the direction of the supervenience intuition, where larger-scale classical phenomena modify the micro-scale quantum phenomena reminiscent of the role of stability conditions in contextual emergence (section 9.4).

While nonlinear dynamics and chaos don't settle debates between these two traditions, chaotic dynamics, its properties, and practices associated with its study may suggest an alternative approach to understanding the regularities in question. Consider Newton's second law, $F = ma$, as an example. Why think the forces, F, 'govern' particle behavior, or of particle behavior as supervening on F? Modeling of forces and system states typically is carried out in terms of energy, constraints, and affordances defined over state spaces delimiting the range of possible system behaviors. Newton's law functions as a constraint on the action of forces along with affording various possibilities for system behaviors (e.g., [24], section 7.8). Among the factors relevant to system behaviors are physical, dimensional (i.e., many forces behave differently in two dimensions versus three), and dynamical (i.e, due to system dynamics arising over time). These different stability conditions define physical contexts for system behaviors. For instance, in fluid convection emergent large-scale fluid convection cells constrain possible motions of smaller-scale fluid molecules under the action of gravity while opening possibilities for other motions not available in the quiescent state [25], section 4.1.

Physical laws, such as the two-way speed of light in vacuum, delineate the bounds of possibility, but they don't determine any concrete actualization of these

---

[6] This tradition is one of the strongest motivations for the governance myth (section 1.9.1).

possibilities. More properties of actual-world contexts are involved in actualization of specific systems and their behaviors.

10.3.3 Chaos and emergence

While the relationship between quantum and classical domains and philosophy of mind are the usual terrains for emergence debates [26, 27], chaos exhibits marks of emergence. For instance, it arises in macroscopic/classical systems with algebras of observables and state spaces distinct from those of QM (chapter 9). Even at macroscopic scales, chaotic dynamics only emerges when parameter magnitudes are in the relevant range, whether considering mathematical models or actual-world systems. This is sensitivity to parameter magnitudes as stability conditions for chaotic dynamics. Likewise, determinism is a relevant stability condition for the emergence of SDIC, aperiodicity, and other signatures of chaos from regular dynamics.

Moreover, as observed in chapter 9, the tools of quantum chaos have no direct correspondence to the tools of classical chaos. The tools and marks of classical chaos don't arise as limiting cases of quantum chaos or as some approximation to the latter. Classical chaos looks to be a quintessential example of emergence, but is this an ontic or merely epistemic phenomenon?

Ontological reductionism is the thesis that properties and behaviors of systems are nothing over and above the states and intrinsic properties of their constituents. On this view, your reading this book is just a particular configuration and play of elementary particles and forces. **Epistemological reductionism** is the related thesis that the states and intrinsic properties of their constituent parts ultimately explain system behaviors. Of course, epistemological reductionism could fail—we could be unable to complete the explanation from elementary particles and forces to the existence and properties of this book and the activity of reading—while believing ontological reductionism holds (e.g., [28, 29]).

Emergence accounts deny these theses yet are often vaguely defined. **Strong** or **radical emergence** is the thesis some kind of brute bridge laws between elementary particles and forces and biological or social phenomena exist. This is the usual ontological contrast with reductionism in the literature but it presents a disunified picture of nature and seems disconnected from the sciences. Strong emergence typically is dismissed in the sciences leaving only epistemological forms of emergence (i.e., failures of epistemological reductionism).

There are alternative ontological accounts of emergence rooted in the sciences and scientific practice for situating chaotic dynamics in a unified understanding. The transformational emergence of Paul Humphreys [30] and Gil Santos [31, 32] highlights dynamical transformations, where smaller-scale properties and constituents often disappear in the formation of new larger-scale properties or entities, or emergent wholes transform the dynamics and properties of smaller-scale properties or entities. Another possibility, contextual emergence, emphasizes multi-scale relations and the roles of stability conditions not found at smaller length and shorter time scales but which nonetheless enable or constrain properties, entities, and processes at these smaller/shorter scales, making modal changes to possibility and actualization at these latter scales [24, 25]. These ontological accounts capture chaos as an emergent phenomenon in intelligible ways exhibiting the order and unity or nature.

The context-dependent nature of chaotic dynamics illustrates contextual emergence (section 9.4). The presence or absence of nonlinearities make chaos possible or impossible, respectively. The magnitude of key factors (parameters in our models) determine when nonlinearities are active. These are examples of stability conditions leading from chaos being a possible to an actual behavior.

10.3.4 Chaos and free will

Troubles for determinism and reductionism lead to speculations about the implications of chaos for free will (e.g., [33]). How relevant and extensive the existence of chaos in brain operations is a hotly debated empirical matter. These discussions typically assume SD or Chaos$_\lambda$ (section 5.2) as definition of chaos, though all that's really needed for sensitivity to and amplification of small changes in the brain is the loss of superposition found in nonlinear systems. Furthermore, as discussed in chapter 7, applying arguments based on SDIC to concrete physical systems is less straightforward than much of the literature takes into account. For instance, we lack knowledge about constraints on or affordances for amplification in neural systems and processes. However, large amounts of parallel redundancy in brain regions and neural networks raise questions about whether changes in one neuron firing differently can be amplified to affect neural dynamics, much less a person's decision.

Moreover, if the macroscopic world is deterministic, invocations of SDIC in cognitive dynamics may provide a sophisticated framework for exploring deliberative processes, but would be insufficient for notions of free will incompatible with determinism. On the other hand, if indeterminism in some form is relevantly operative in the brain, incompatibilists, such Robert Kane [33], face challenges demonstrating agents can effectively harness such indeterminism through the exquisite sensitivity provided by nonlinear dynamics to explain free will.[7]

10.3.5 Divine action in a nonlinear world

Finally, some, such as physicist turned Anglican priest John Polkinghorne, argue that 'SDIC is an indication of nature's ontological openness due to chaos and complex dynamical systems' (1989, p 43). Following a critical realist approach to epistemology [35], he links the epistemic failure of determinism—apparent randomness (section 1.1)—with an ontic failure of determinism. The latter failure of unique evolution, he argues, is due to an ontological openness to influences not fully accounted for in physics descriptions. One could make similar arguments drawing on issues with the faithful model framework (section 1.9), raising questions about the extent to which our deterministic models of macroscopic systems inadequately

---

[7] For instance, see the exchange between Kane and Mauro Dorato in [34].

characterize actual-world systems or miss out influences outside the descriptive apparatus of physics.

In this sense, Polkinghorne proposes interpreting randomness in macroscopic chaotic systems as representing genuine indeterminism rather than being merely apparent [36]. He argues such openness isn't only important to human free will, but also for divine action in the world (e.g., [37]). The basic idea is that if SDIC possibly opens the door to quantum influences, then it opens the door to other macroscopic influences beyond physics, too (e.g., biological, psychological). In analogy, SDIC opens the door for divine influence in nature. Polkinghorne focuses more on possible manipulations of quantum events amplified by SDIC as the route for divine action in nature.

This line of reasoning clearly assumes the SD argument discussed in chapter 7, which, as we have seen, is problematic. Moreover, it assumes nature is a closed deterministic system, where God needs 'cracks' or 'openings' via some form of indeterminism to act in the world. This is an example of an interventionist approach to divine action, where God actively modifies some physical circumstance to change an outcome in nature, in Polkinghorne's case some small physical change subsequently amplified by chaotic dynamics.

Alternative approaches maintain a strict Creator-creature distinction but emphasize divine action working through the properties and processes of nature without such interventions (e.g., [38]). In these latter approaches, both microscopic and macroscopic processes, including forms of SDIC, represent a means by which change can occur as well as contingency for nature's properties and processes to act with integrity requiring little to no direct divine intervention in natural processes (e.g., [39, 40]). Such approaches are more consistent with a coherent, unified understanding of nature, where nonlinearity plays surprising roles as illustrated in this book.

10.4 Prospects

As this closing chapter indicates, chaos holds much promise in terms of the richness of its behavior in our world and the insights it can provide from the interior of cells, to the Solar System, and beyond. Meanwhile, chaos also raises questions about forecasting limitations, how much we should expect our world to be deterministic, what its causal structure is, and what limits there might be to obtaining answers to some of our questions.

Studying chaotic dynamics over the decades has taught as much about modeling, how to use computers well, how to interpret 'noise' in systems, and how to face uncertainty. Chaos has been a useful teacher for scientists exploring the world and its order and likely has more surprises for us to learn as we continue those explorations.

Finally, chaos studies have taught us lessons about living with limits. From forecasting limits, to limits on measurement accuracy, to working with uncertainty, scientists have grown in their understanding of nature and how to manage the limits she places on us. Far from being pessimistic about the challenges chaotic dynamics reveals, we have learned more about nature and have become more sophisticated and adept at how we cope with and apply chaos and the uncertainty it highlights. There is no reason to think this trend will end anytime soon.

References

[1] Woodman S M *et al* 2019 ESDM: a tool for creating and exploring ensembles of predictions from species distribution and abundance models *Methods Ecol. Evol.* **10** 1923–33

[2] Kranstauber B *et al* 2022 Ensemble predictions are essential for accurate bird migration forecasts for conservation and flight safety *Ecol. Solut. Evid.* **3** e12158

[3] Munch S B *et al* 2022 Rethinking the prevalence and relevance of chaos in ecology *Annu. Rev. Ecol. Evol. S* **53** 227–49

[4] Newton I 1730 *Opticks: Or, a Treatise of the Reflections, Refractions, Inflections and Colours of Light* ed 4th corrected (London: William Innys)

[5] Mogavero F *et al* 2023 Timescales of chaos in the inner solar system: Lyapunov spectrum and quasi-integrals of motion *Phys. Rev. X* **13** 021018

[6] Adkins R *et al* 2022 Dynamics of active liquid interfaces *Science* **377** 768–72

[7] Panahi S *et al* 2019 A new chaotic network model for epilepsy *Appl. Math. Comput.* **346** 395–407

[8] Karavaev A S *et al* 2019 Autonomic control is a source of dynamical chaos in the cardiovascular system *Chaos* **29** 121101

[9] Singh R S *et al* 2021 Automated detection of normal and cardiac heart disease using chaos attributes and online sequential extreme learning machine *Computational Intelligence in Healthcare* ed A K Manocha *et al* (Cham: Springer) 213–34 pp

[10] Uthamacumara A and Zenil H 2022 A review of mathematical and computational methods in cancer dynamics *Front. Oncol.* **12** 850731

[11] Minati L *et al* 2022 Incomplete synchronization of chaos under frequency-limited coupling: observations in single-transistor microwave oscillators *Chaos Solitons Fractals* **165** 112854

[12] Zaher A A and Abu-Rezq A 2011 On the design of chaos-based secure communication systems *Commun. Nonlinear Sci. Numer. Simul.* **16** 3721–37

[13] Skiadas C H and Skiadas C (ed) 2016 *Handbook of Applications of Chaos Theory* (Boca Raton, FL: CRC Press) https://doi.org/10.1201/b20232

[14] Bishop R C 2002 Deterministic and indeterministic descriptions *Between Chance and Choice: Interdisciplinary Perspectives on Determinism* ed H Atmanspacher and R Bishop (Thorverton: Imprint Academic) 5–31 pp

[15] Bishop R C 2003 On separating prediction from determinism *Erkenntnis* **58** 169–88

[16] Chirikov B V 1979 The problem of quantum chaos *Chaos and Quantum Chaos: Proc. 8th Chris Engelbrecht Summer School on Theoretical Physics*; W D Heiss (Berlin: Springer) 1–56

[17] Bishop R C 2011 Metaphysical and epistemological issues in complex systems *Philosophy of Complex Systems, vol 10 of Handbook of the Philosophy of Science* ed C Hooker (Amsterdam: Elsevier) 105–36

[18] Hadamard J 1922 *Lectures on Cauch's Problem in Linear Partial Differential equations* (New Haven, CT: Yale University Press)

[19] Arnold V I 1988 Geometrical Methods in the Theory of Ordinary *Differential equations.* (New York: Springer) https://doi.org/10.1007/978-1-4612-1037-5

[20] Bishop R C and beim Graben P 2016 Contextual emergence of deterministic and stochastic descriptions *From Chemistry to Consciousness: The Legacy of Hans Primas* ed H Atmanspacher and U Müller-Herold (Cham: Springer) 95–110

[21] Oakley F 1961 Christian theology and the Newtonian science: the rise of the concept of the laws of nature *Church Hist* **30** 433–57

[22] Kellert S H 1993 *In the Wake of Chaos* (Chicago, IL: University of Chicago Press)

[23] Carroll J W 2024 Laws of nature *The Stanford Encyclopedia of Philosophy* (Winter 2024 Edition), E N Zalta and U Nodelman (eds.), https://plato.stanford.edu/archives/win2024/entries/laws-of-nature/

[24] Bishop R C *et al* 2022 *Emergence in Context: A Treatise in Twenty-First Century Natural Philosophy* (Oxford: Oxford University Press) https://doi.org/10.1093/oso/9780192849786.001.0001

[25] Bishop R C 2024 *The Physics of Emergence* 2nd ed (Bristol: IOP Publishing) https://doi.org/10.1088/978-0-7503-6367-9

[26] Bedau M and Humphreys P (ed) 2008 *Emergence: Contemporary Readings in Philosophy and Science* (Cambridge, MA: MIT Press) https://doi.org/10.7551/mitpress/9780262026215.001.0001

[27] Gibb S *et al* (ed) 2019 *Routledge Handbook of Emergence* (Abingdon: Routledge) https://doi.org/10.4324/9781315675213

[28] Anderson P W 1972 More is different: broken symmetry and the nature of the hierarchical structure of science *Science* **177** 383–96

[29] Laughlin R B A 2005 *A Different Universe: Reinventing Physics from the Bottom Down* (New York: Basic Books)

[30] Humphreys P 2016 *Emergence: A Philosophical Account* (Oxford: Oxford University Press) https://doi.org/10.1093/acprof:oso/9780190620325.001.0001

[31] Santos G C 2015 Ontological emergence: how is that possible? Towards a new relational ontology *Found. Sci.* **20** 429–46

[32] Santos G C 2020 Integrated-structure emergence and its mechanistic explanation *Synthese* **198** 8687–711

[33] Kane R 1998 *The Significance of Free Will* (Oxford: Oxford University Press)

[34] Atmanspacher H and Bishop R (ed) 2002 *Between Chance and Choice: Interdisciplinary Perspectives* (Thorverton: Imprint Academic)

[35] Andrew C 1998 Critical realism *Routledge Encyclopedia of Philosophy* (London: Taylor and Francis) https://www.rep.Routledge.com/articles/thematic/critical-realism/v-1, https://doi.org/10.4324/9780415249126-R003-1

[36] Reason P J 1991 *Reason and Reality: The Relationship* (Valley Forge, PA: Trinity Press)

[37] Polkinghorne J 1991 *Science and Creation: The Search for* (Boston, MA: Shambhala Publications)

[38] Gunton C E 1998 *The Triune Creator: A Historical and Study* (Grand Rapids, MI: Eerdmans)

[39] Bishop R C *et al* 2018 *Understanding Scientific Theories of Origins: Cosmology, Geology, and Biology in Christian Perspective* (Downers Grove, IL: IVP Academic)

[40] Koperski J 2020 The Physics of Theism: God *Physics, and the Philosophy of Science* (Chichester: Wiley)

IOP Publishing

An Introduction to Chaotic Dynamics
Classical and quantum
Robert C Bishop

Appendix A

The chaotic hierarchy

Some have thought it possible to define chaos as a hierarchy from weak to strong based on the ergodic hierarchy (e.g., [1, 2]). The ergodic hierarchy is an ordering of dynamical system properties: ergodicity, weak mixing, strong mixing, K-systems, and B-systems.

Definition A.1. Ergodicity: starting from almost any initial condition, a single trajectory on an energy surface will explore the entire available state space densely. For all initial conditions except a set of measure zero, the time average and the space average of the function describing the trajectory on an energy surface are equal.

Ergodicity is at the base of the hierarchy and is the least mathematical of the definitions.

Assume a state space S is equipped with a measure μ and let T_μ be a measure preserving transformation such as a map or a flow and consider subsets A and B of S. The remaining properties in the hierarchy can be defined in order as follows:

Definition A.2. Weak mixing: given two sets of points A and B of an energy surface in S (excluding those of zero measure), $\lim_{n\to\infty} \frac{1}{n}\sum_{i=0}^{n-1}|T_\mu^{-i}A \cap B - \mu(A)\mu(B)| = 0$.

Definition A.3. Strong mixing: given two sets of points A and B of an energy surface in S (excluding those of zero measure), $\lim_{n\to\infty}(T_\mu^{-i}A) \cap B - \mu(A)\mu(B) = 0$.

Definition A.4. K-systems (Kolmogorov): quasi-regular dynamical systems having positive Kolmogorov entropy, also known as Kolmogorov–Sinai entropy, where $k(T, \xi) = (T_\mu^{-1}A_{i_1} \cap T_\mu^{-2}A_{i_2} \cap T_\mu^{-3}A_{i_3} \cap \cdots \cap T_\mu^{-n}A_{i_n})$. Let

doi:10.1088/978-0-7503-6453-9ch11
A-1

$$h(T, \xi) = \lim_{n \to \infty} \frac{1}{n} \sum_{i_1, i_2, i_3, \ldots, i_n} k(T, \xi) \ln k(T, \xi) \qquad (A.1)$$

be the entropy over the finite partition $\xi = \{A_{i_1}, A_{i_2}, A_{i_3}, \ldots, A_{i_n}\}$ of S. A dynamical system is a K-system if $h(T) = \sup_{\xi} h(T, \xi)$, where the supremum is taken over all finite partitions ξ.

Definition A.5. B-systems (Bernoulli): dynamical systems possessing a generating partition for T such that for all i, $\bigvee_{j=1}^{i-1} T^j \xi$ and $T^j \xi$ are independent, where $A \bigvee B = \{A_i \cap B_j\}$.

The ordering of the hierarchy is such that any dynamical systems that are B-system also are K-system. Any systems that are K-system also are strong mixing. Any systems that are strong mixing also are weak mixing. Any weak mixing systems are ergodic. On the other hand, merely ergodic systems share none of the other properties of the hierarchy, weak mixing systems are ergodic but lack strong mixing, and so forth.[1]

Sometimes the ergodic hierarchy is interpreted as representing different degrees of randomness (e.g., [3]). Nevertheless, note that any degrees of randomness in the hierarchy are only apparently random (section 1.1) since all dynamical systems are deterministic.

Applying this classification scheme for dynamical systems to chaos raises subtleties. For instance, dissipative systems may have an attractor that is ergodic only in a limited region of S implying a trajectory only densely covers the attractor. In conservative systems, chaotic trajectories often don't cover the entire state space densely. The Chirikov standard map (section 3.2) is an example where ergodicity could only hold for regions of S where there is no mixture of both chaotic and regular orbits. As a practical matter, chaotic models are run on computers, where the graphical output can look convincing for ergodicity to the eye, but the discrete nature of the calculations implies that we cannot rigorously demonstrate ergodicity.

One approach for constructing a chaotic hierarchy is to associate a form of 'weak chaos' with systems exhibiting ergodicity or some form of mixing, while associating 'strong chaos' with systems exhibiting SDIC and/or continuous energy spectra, loss of temporal correlations, and other properties that appear in K- and B-systems (e.g., [2, 4]). The intuition is that there is a gradation of levels chaotic dynamics. For example, ergodicity produces WSD but not SD.

Another approach is to identify strong mixing as a necessary condition for chaos while being a K-system is sufficient (e.g., [5]) because only K-systems exhibit SDIC on the ergodic hierarchy. This is equivalent to a positive Kolmogorov–Sinai entropy being necessary and sufficient for a positive global Lyapunov exponent (e.g., [6]), but as discussed in chapter 5 and appendix C, such exponents are of limited pragmatic

---

[1] Some authors equate ergodicity with weak mixing.

A-2

value in distinguishing chaotic from other kinds of dynamics. Charlotte Werndl [7] has argued that strong mixing implies that trajectories will cover the state space (or an attractor in state space) densely, implying Chaos$_D$. However, as discussed in chapter 5, Chaos$_D$ is neither necessary nor sufficient to reliably distinguish chaotic behavior.

Since randomness is connected to unpredictability, another approach to constructing a chaotic hierarchy starts with weak mixing. Proceeding up successive levels of the hierarchy, a dynamical system's past behavior becomes less and less effective (some might argue less and less relevant) to predicting future behavior (e.g., [8]). However, we currently lack a rigorous measure distinguishing a specific degree of unpredictability unequivocally associated with chaotic dynamics.

Furthermore, unpredictability of chaotic dynamics is more subtle than the ergodic hierarchy can take into account. Scientists engaged in forecasting always deal with finite uncertainties and their growth, yet typically don't bother computing global Lyapunov exponents. For instance with weather forecasting, scientists use observations of the initial state of their system yielding ensembles of initial conditions. They then assimilate this data into the model state space, forming the initial conditions for the forecasting model and use ensemble methods to produce weather forecasts with rather remarkable accuracy (chapter 6).

Chaotic dynamics is present in weather systems. Nevertheless, scientists have techniques for coaxing good, useful forecasting for such systems over the near and medium-range future. Chaotic dynamics doesn't render your daily weather forecast useless because of unpredictability. Similarly for many other actual-world chaotic systems. Unpredictability doesn't appear to to yield a chaotic hierarchy.

References

[1] Zhilinskiĭ B I 2009 Quantum bifurcations *Encyclopedia of Complexity and Systems Science* ed R Meyers (New York: Springer) pp 7135–54

[2] Ullmo D 2016 Bohigas-Giannoni-Schmit conjecture *Scholarpedia* **11** 31721

[3] Ott E 2002 *Chaos in Dynamical Systems* 2nd ed (Cambridge: Cambridge University Press) https://doi.org/10.1017/CBO9780511803260

[4] Casati G and Prosen T 2009 *Quantum chaos Encyclopedia of Complexity and Systems Science* (New York: Springer) 7164–74

[5] Belot G and Earman J 1997 Chaos out of order: quantum mechanics, the correspondence principle and chaos *Stud. Hist. Phil. Mod. Sci.* **28** 147–82

[6] Lichtenberg A J and Liebermann M A 1992 *Regular and Chaotic Dynamics* 2nd ed (New York: Springer) https://doi.org/10.1007/978-1-4757-2184-3

[7] Werndl C 2009 What are the new implications of chaos for unpredictability? *Br. J. Phil. Sci.* **60** 195–220

[8] Berkovitz J *et al* 2006 The ergodic hierarchy, randomness and chaos *Stud. Hist. Phil. Mod. Phys.* **37** 661–91

IOP Publishing

An Introduction to Chaotic Dynamics
Classical and quantum
Robert C Bishop

Appendix B

Topological entropy

Topological dynamics is the study of long term or asymptotic properties of various kinds of transformations of topological spaces. Consider a topological state space S and a topological group transformation T. Abstract topological dynamics is defined for a flow (S, T) under the jointly continuous action of T on S: there is a continuous transformation, $T \times S \to S$, with $(t, x) \to tx$, where $t \in T$ is time, x a point in S, $t_1(t_2 x) = (t_1 t_2)x$, and the identity $I \in T$, $Ix = x$.

If S is a metrizable topological space, and (S, T) a topological dynamical system, then the **topological entropy** can be defined as a nonnegative number measuring the complexity of the system, the exponential growth rate of the number of distinguishable orbits as time advances: let the set $E \subset S$ be (n, ε)-separated, in other words for every $x, y \in E$ with $x \neq y$ there is a $j \in \{0, 1, 2, \ldots, n - 1\}$ such that $d(T^j x, T^j y) \geqslant \varepsilon$, where d is the metric.. Let $\alpha(n, \varepsilon)$ be the maximal cardinality of an (n, ε)-separated set in S. The the quantity $\bar{h}$ can be defined as

$$\bar{h}(\varepsilon, T) = \limsup_{n \to \infty} \frac{\log \alpha(n, \varepsilon)}{n}.$$

Then the topological entropy can be defined as

$$h_B(T) = \lim_{\varepsilon \to 0} \bar{h}(\varepsilon, T). \tag{B.1}$$

The interpretation of $\alpha(n, \varepsilon)$ is that two points x and y are distinguished only if the distance between them is at least ε. After the dynamical system evolves n steps, at most only $\alpha(n, \varepsilon)$ orbits are distinguishable. This implies that $\bar{h}(\varepsilon, T)$ is the exponential growth rate of the number of ε-distinguishable n-orbits as $n \to \infty$. The value of $h_B(T)$ maximizes this over all $\varepsilon > 0$. This is the sense in which topological entropy measures the exponential complexity of a dynamical system.

doi:10.1088/978-0-7503-6453-9ch12   B-1   © IOP Publishing Ltd 2025. All rights,

Appendix C

Global Lyapunov exponents

Global Lyapunov exponents arise out of linear stability analysis of the trajectories of evolution equations such as (1.1)–(1.3). Consider the first-order, ordinary differential equation system $d\vec{x}/dt = F\vec{x}$. Let $\vec{x}^*$ a fixed point where $F(\vec{x}^*) = 0$. Further, let $\vec{x}(t) = \vec{x}^* + \varepsilon(t)$, where $\varepsilon(t)$ is an infinitesimal perturbation to every component of $\vec{x}^*$. A linear stability analysis involves substituting back into F and expanding to first order in $\varepsilon(t)$ while considering only the perturbations at $t = 0$ (dropping the explicit t-dependence from ε):

$$F(\vec{x}^* + \varepsilon) = F(\vec{x}^*) + J(\vec{x}^*)\varepsilon + O(|\varepsilon|^2), \qquad (C.1)$$

where $J(\vec{x}^*)$ is the propagator of F evaluated at $\vec{x}$. This yields an equation for the time dependence of the perturbation of x:

$$\frac{d\varepsilon}{dt} = J(\vec{x}^*)\varepsilon + O(|\varepsilon|^2). \qquad (C.2)$$

Neglecting all terms $O(|\varepsilon|^2)$ and higher in equation (C.2) leads to a linear stability analysis for the perturbation. Assume ε is real and J a real-valued matrix. A solution of the form $\varepsilon = \lambda_i e^{at}$ yields the eigenvalue equation

$$J\lambda = a\lambda, \qquad (C.3)$$

where λ is a column vector. In general, there are n components corresponding to an n-dimensional system. Consider the initial condition $\vec{x}(0)$ and an infinitesimal displacement from $\vec{x}(0)$ in the direction of some tangent vector, $\vec{y}(0)$. Then the evolution of $\vec{y}$ is given by

$$\frac{d\vec{y}}{dt} = J(\vec{x})\vec{y}, \qquad (C.4)$$

valid for an infinitesimal neighborhood about $\vec{x}(0)$. The value of $\vec{y}(t)$ changes with time, where $\vec{y}/|\vec{y}|$ gives the direction the infinitesimal displacement from $\vec{x}$ while

doi:10.1088/978-0-7503-6453-9ch13

$\vec{y}/|\vec{y}(0)|$ gives the magnitude by which the infinitesimal displacement grows ($|\vec{y}| > |\vec{y}(0)|$) or shrinks ($|\vec{y}| < |\vec{y}(0)|$).

The **global Lyapunov exponent** is now defined with respect to $\vec{x}(0)$ and initial orientation of the infinitesimal displacement $\vec{y}(0)/|\vec{y}(0)|$ as

$$\lambda(\vec{x}(0), \vec{y}/|\vec{y}(0)|) = \lim_{t \to \infty} \frac{1}{t} \ln\left(\frac{d\vec{y}/dt}{\vec{y}(0)}\right) = \lim_{t \to \infty} \frac{1}{t} \ln\left(\frac{J(\vec{x})\vec{y}}{\vec{y}(0)}\right). \tag{C.5}$$

The relevant global Lyapunov exponent is distinguished by the initial orientation of the perturbation. Note that the infinite time limit plays an important role in this analysis, implying Lyapunov exponents represent time-averaged quantities (i.e., transient behavior has decayed). The existence of this limit is guaranteed by Oseledets' multiplicative ergodic theorem [1], which holds under mild conditions. Since J is a constant in space in this limit, global Lyapunov exponents obtained from (C.5) are then the same for almost every value of $\vec{x}(0)$.

Global Lyapunov exponents are calculated in one of two ways. The first approach is to begin by selecting two points in state space separated by a small finite distance d_0, advance the system of equations one time step, then recompute the distance d between the two trajectories and take the logarithm of this distance. These steps are repeated for as many time steps as needed. The results are then averaged since the distance between the trajectories might diverge from each other on one time step, converge on another time step, and so on. This approach assumes that the state space has a metric so distances can be measured. Every few time steps the distance must be recalibrated to a new d_0 accounting for how the spreading between the two trajectories is expanding and contracting on the attractor.

This approach has issues. For example, if the starting conditions are too close to a limit cycle, accurate global Lyapunov components cannot be calculated. Furthermore, if the two initial points are too close to each other, then trajectory spread can be modeled with any exponent (i.e., there may be no unique exponent to identify with a global Lyapunov exponent).

The approach of choice for computing global Lyapunov exponents are variational methods. A variational equation is evolved for the growth rate of the orthogonal initial deviation vectors of an ellipse centered on $\vec{x}(0)$. Although more complicated than indicated by this brief summary, variational approaches avoid the need to repeatedly renormalize distance between two trajectories.

Reference

[1] Oseledets V I 1968 A multiplicative ergodic theorem: Lyapunov characteristic numbers for dynamical systems *Trans. Moscow Math. Soc.* **19** 179–210

www.ingramcontent.com/pod-product-compliance
Lightning Source LLC
Chambersburg PA
CBHW062009190326
41458CB00009B/3016